Smart Antennas and Signal Processing
for Communications, Biomedical and Radar Systems

Smart Antennas and Signal Processing

for Communications, Biomedical and Radar Systems

Editor:

P.R.P. Hoole, D.Phil. Oxon.

Nanyang Technological University

WITPRESS Southampton, Boston

Editor:

P.R.P. Hoole
Nangyang Technological University, Singapore

Published by

WIT Press
Ashurst Lodge, Ashurst, Southampton, SO40 7AA, UK
Tel: 44 (0) 238 029 3223; Fax: 44 (0) 238 029 2853
E-Mail: witpress@witpress.com
http://www.witpress.com

For USA, Canada and Mexico

Computational Mechanics Inc
25 Bridge Street, Billerica, MA 01821, USA
Tel: 978 667 5841; Fax: 978 667 7582
E-Mail: info@compmech.com
US site: http://www.compmech.com

British Library Cataloguing-in-Publication Data

A Catalogue record for this book is available
from the British Library

ISBN: 1-85312-801-5

Library of Congress Catalog Card Number: 99-68743

*The texts of the papers in this volume were set
individually by the authors or under their supervision.
Only minor corrections to the text may have been carried
out by the publisher.*

Printed in Great Britain by IBT Global Ltd., London.

Contents

Chapter 3. Focused beam antennas
P.R.P. Hoole

Chapter 4. Antenna beamforming
P.R.P. Hoole

Chapter 5. Antennas and signals in magnetic resonance imaging
T.S. Naveendra and P.R.P. Hoole

Chapter 6. Synthetic aperture antennas and imaging
Tan Pek Hua, Dennis Goh, P.R.P. Hoole and U.R. Abeyratne

Chapter 7. Smart antennas: mobile station antenna location
Stetson Oh Kok Leong, Ng Kim Chong, Paul R.P. Hoole and E. Gunawan

Chapter 8. Smart antennas: mobile station antenna beamforming
Ng Kim Chong, Stetson Oh Kok Leong, Paul R.P. Hoole and E. Gunawan

Chapter 9. Smart antennas: base station antenna beamforming
Ng Yin Yoon, Edmund Yap Aik Boon and Paul R.P. Hoole

Software programs to accompany the following chapters are contained on the disk:

Chapter 2: Program 2.1 Program listings
Chapter 3: Program 3.1 MATLAB^{TM} program listings
Chapter 4: Program 4.1 Programs for the Woodward–Lawson sampling method
for both the line source and the linear array
Program 4.2 Program for adaptive antenna
Chapter 5: Program 5.1 MATLAB^{TM} programs I
Program 5.2 MATLAB^{TM} programs II
Program 5.3 MATLAB^{TM} programs III
Program 5.4 MATLAB^{TM} programs IV
Chapter 6: Program 6.1 MATLAB^{TM} program file used for simulation
Program 6.2 MATLAB^{TM} program file used for simulation
Chapter 7: Program 7.1 Program for mobile station position estimation
Chapter 8: Program 8.1 Program for mobile station adaptive beamforming
Chapter 9: Program 9.1 Program listing

Dedication

To the memory of my parents Reverend Richard Hoole and Mrs. Jeevamany Hoole, and to my wife Chrishanthy, and children Esther and Ezekiel. For the love, faithfulness and support of the Triune God that I have known through them.

Sole gratia, et soli Deo gloria.

Foreword

This book provides a clear and simple treatment of antenna theory and applications, with an added emphasis on signal processing and imaging. The first four chapters of the book cover the essential and fundamental theory of all types of antennas at a senior undergraduate level. The final four chapters deal with advanced topics, including the use of array antennas for imaging and the processing of mobile communication antenna signals. The book grew out of the author's many years of teaching and research on antennas and related subjects.

To the best of our knowledge, this is the only book that covers in reasonable depth both traditional antenna theory, antennas in imaging systems and the most recent developments in adaptive antenna technology. Antennas are no longer thought of as stand-alone devices, but have incorporated into them signal processing and beamforming algorithms that make them very powerful, active devices used in communication and imaging systems. The book seeks to capture these overall developments and exciting features of modern antenna technology through a theoretical and practical approach. A common thread that unifies the book is the concept of antenna as a temporal–spatial filter, a device that is capable of operating on signals in the time and spatial domains.

Many useful computer program listings are given so that the students may actually see how antennas work, learn how to develop signal processing modules for antennas and understand and visualize the design and implementation of smart antennas and complex (e.g. MRI and ISAR) imaging routines using the concept of encoded antenna signals and synthetic antennas. It is our view that, when the engineering student sees numbers associated with equations, he or she becomes more confident about the underlying theory as well as its use in practical situations. The computer program listings given in the book are meant only for beginners, and are not intended for professional or commercial use.

With the rapid increase in satellite, mobile and remote sensing systems, properly designed antennas would greatly increase the effectiveness of the systems, in addition to increasing the efficiency of the use of the frequency spectrum. Smaller antennas are used as probes in biomedical applications and other diagnostic systems, in addition to their wide use in mobile and military communications. This book aims to provide a clear, in-depth and concise description of antenna analysis and design. Beginning with the most commonly used wire antennas, a concise treatment of the aperture and array antennas is presented. Antenna synthesis, or the solution of the inverse problem, is introduced by looking at the antenna as a spatial filter. The antenna is not a stand-alone device. Its completeness is tied to the processing of the signals picked up by the antenna. In the final two chapters the book provides a description of the important algorithms which process the signals received by the antenna in telecommunications, radar and biomedical systems. Although the emphasis is on the theoretical understanding and design of antennas, the book contains practical insight into antenna troubleshooting, selection and installation. The book will be of value to the undergraduate student, the graduate student and the practising engineer.

The motivation for writing this book arose from the need that the authors felt for a text on antennas which would cover all the important aspects related to antennas in an attractive, easy-to-digest and practical manner. The audiences addressed are undergraduate students and graduate students just getting into the demands of problem formulation and solution in antennas and related subjects. There are many textbooks on electromagnetic fields that give a brief introduction to antennas. There are other books that go into great details and analysis of antennas. Even those books that give a detailed analysis of antennas often fail to interest the student by missing out the important engineering details related to antennas. Very few books cover the signal processing aspects of antenna design and system engineering. Furthermore, books that are generally addressed to practising engineers tend to play down the importance of the theoretical foundation that is necessary for better design and troubleshooting of antenna associated systems. This book makes an attempt to fill in the gap with a small useful volume. A special word of thanks to Miss Clare Garcia (Production Editor), Mr. Lance Sucharov (Managing Director) and other members of the staff of WIT Press who have rendered my books incomparable services and given me patient assistance. With my wife I would like to acknowledge the volumes of affection and strength both our families have been to us in all our endeavors together.

"The religious feeling of the scientist takes the form of rapturous amazement at the harmony of the natural law which reveals an intelligence of such superiority that, compared with it, all systematic thinking and acting of human beings is an utterly insignificant reflection."

– Albert Einstein

1 Introduction

P.R.P. Hoole

1.1 Elementary principle

An antenna is a device that transmits and receives electromagnetic waves or signals. When transmitting it has to send the signals in a particular direction or in all directions. Radiation must be maximized in a given direction or in some cases omni-directionally. The transmitting and receiving patterns of any antenna are identical. Figure 1.1 shows the transmitting and receiving radiation patterns of an antenna.

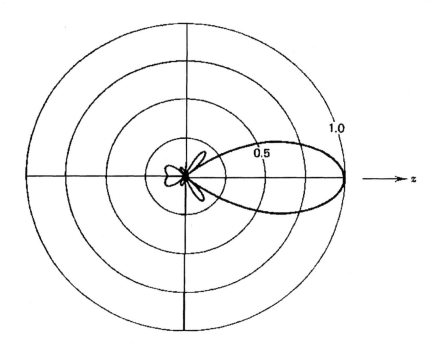

Figure 1.1: Radiation pattern of an antenna.

The radiation pattern indicates the magnitude of electric field strength observed at a constant distance r from the antenna in different directions. Most antennas, as indicated, are designed to transmit and receive maximum signal in one particular direction at a specified center frequency and bandwidth.

How does an antenna work? The principle of antenna operation may be best explained by considering Faraday's law, which defines the induced voltage V in a coil as given by

$$V = \frac{\mathrm{d}(N\phi)}{\mathrm{d}t} = \frac{\mathrm{d}(NBA)}{\mathrm{d}t}$$

$$= N\left[B\frac{\mathrm{d}A}{\mathrm{d}t} + A\frac{\mathrm{d}B}{\mathrm{d}t}\right], \tag{1}$$

where N is the number of turns, B is the flux density and A is the area of the coil. We note that there are two terms to the induced voltage. A receiving antenna may make use of one or both these terms. When a time varying magnetic field cuts a conductor, it will induce a voltage, which will be proportional to $\mathrm{d}B/\mathrm{d}t$. This is how a wire antenna of a mobile telephone picks up signals. Alternatively, if a coil antenna is rotated in an electromagnetic field, there will be an induced voltage due to $\mathrm{d}A/\mathrm{d}t$, where the area of the coil cutting the signal is changing. Such antennas are used for direction finding; when the induced voltage is maximum, the antenna is directly facing the source that is emitting the electromagnetic signal. In military applications, such rotating antennas may be used to detect hidden transmitters and radar. In a transmitting antenna, a wire to which a time varying voltage (e.g. an FM radio signal) is applied will produce time varying magnetic flux density B in the space around it. At high frequencies, this B will produce an electric field intensity E. Both E and B together will now propagate through free space at the velocity of light.

1.2   Broadcast frequency bands

When we speak of frequency bands, we are describing the frequencies used in radio, TV and other forms of electronic communications. These frequencies start at 1 Hz, then progress through the audio frequencies of 20–20,000 Hz, then into the nine radio bands, all the way into the frequency range of visible light. From 1 to 300 GHz is generally considered to be microwave. The historic frequency band designations are as follows: extra-low-frequency, ELF (0–300 Hz); very-low-frequency, VLF (3–30 kHz); low-frequency, LF (30–300 kHz); high-frequency, HF (0.003–0.03 GHz); very-high-frequency, VHF (0.03–0.3 GHz); ultra-high-frequency, UHF (0.3–3.0 GHz); super-high-frequency, SHF (3–30 GHz); extra-high-frequency, EHF (30–300 GHz); L-band (1.0–2.0 GHz), S-band (2.0–4.0 GHz), C-band (4.0–8.0 GHz), X-band (8.0–12.5 GHz), Ku-band (12.5–18.0 GHz), K-band (18.0–26.5 GHz) and Ka-band (26.5–40.0 GHz). The millimeter (mm) waves range from 300 to 3000 GHz. There are new band designations too that are simpler than the historic designations: A (0–0.25 GHz), B (0.25–0.5 GHz), C (0.5–1.0 GHz), D (1.0–2.0 GHz), E (2.0–3.0 GHz), F (3.0–4.0 GHz), G (4.0–6.0 GHz), H (6.0–8.0 GHz), I (8.0–10.0 GHz), J (10.0–20.0 GHz) and K (20.0–40.0 GHz). We shall mostly use the historic band designations.

Communications, imaging and radar antennas are normally designed for operation in the radio, TV and microwave frequency bands. Consider a digital communication system. All image, video and electronic mail data are represented by bits 1 and 0. For long distance communications, these rectangular pulses are modulated into sinusoidal signals. In frequency modulation, bit 1 may be represented by frequency f_1, and bit 0 by frequency f_2. In phase modulation, bit 1 may be represented by a sinusoidal signal at frequency f at phase 0°, whereas bit 0 may be represented by a sinusoidal signal at the same frequency but at phase 45°. Thus in either case, the signal to be handled by the antenna is a sinusoidal electromagnetic signal. Now to pack more data (i.e. more 1s and 0s) into a short wave packet, higher frequencies have to be used, since one bit typically takes $1/f$ s. This is the reason why communication systems carrying large amounts of information (e.g. two TV channels and 3000 telephone channels) must use the microwave frequency bands. However, low-data-rate AM and FM radio stations use the lower frequency spectra. Antennas must be able to handle all these frequency bands. The type and size of antenna will depend on its frequency of operation. Infrared (wavelength range 10^{-4}–10^{-6} m) used in night vision, visible light (wavelength range 720×10^{-9} m for deep red to 380×10^{-9} m for violet), ultraviolet light (wavelength range 10^{-7}–10^{-11} m used in sterilization), X-rays (wavelength range 10^{-10}–10^{-14} m) used in medical diagnostics like CT scans, and γ-rays (wavelength range 10^{-11}–10^{-16} m) need other kinds of radiators which are not dealt with here.

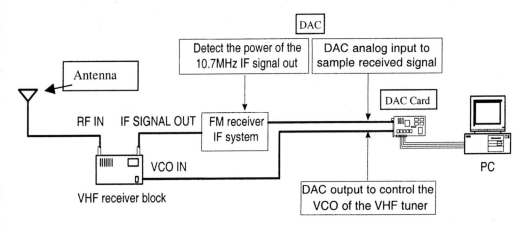

Figure 1.2: A communications receiver system.

Figure 1.2 shows a simple communications receiver system and the electromagnetic spectrum. The particular VHF receiver shown in Fig. 1.2 is designed to work at a frequency of about 200 MHz. The form in which data is received is the analog sinusoidal waveform. The 200 MHz waveform must be frequency downconverted (by the VHF receiver), processed (FM receiver) and converted into digital form by the digital-to-analog converter (DAC) before being stored in the converted form in a personal computer (PC). Such will be the sequence of steps adapted by an e-mail (electronic mail) received via a wireless system into your PC. The DAC is necessary only if the final form of the data should be in digital form, as in the case of e-mails. In the whole system it is only in the final stage after the DAC that the waveform becomes a non-sinusoidal waveform of square pulse shape (e.g. 5 V for bit 1 and 0 V for bit 0).

Consider the issue of frequency bandwidth. A calling card is a credit card which enables you to make a telephone call from a public booth, and to charge it on your private or office telephone. Note the enormous number of digits required. First you dial the number you want to call (e.g. 298 1120) at area code, e.g. 818. Then your private telephone number (e.g. 7780 7698) to which you want the charge to be assigned.

Each number that you press is assigned two tones as indicated in Table 1.1. For instance, if you press 2, then two electrical signals at frequencies 1336 and 697 Hz are transmitted along the telephone wire to the central exchange station. Then if it is wireless, both frequencies will be upconverted to frequencies around the carrier frequency of the wireless system (e.g. 900 or 1980 MHz). Thus each number on dialing occupies two frequencies on the frequency spectrum. For the complete set of numbers, seven frequencies are required. Now this is only for the digits. When you start talking, more frequencies are required (if it is an analog FM system) to represent your voice tones. Hence each voice link occupies a certain frequency bandwidth on the frequency spectrum. An antenna designed for a specific system (e.g. a mobile communication link at a carrier frequency of 800 MHz and a bandwidth of 100 kHz) will have to effectively launch and direct signals at that particular carrier frequency and over the specified bandwidth. Table 1.1 shows the seven push button frequencies of a telephone unit. When one button is pressed, it will generate a signal that is made of a combination of two frequencies.

Table 1.1: Frequencies assigned to buttons on a telephone unit.

	1209 Hz	1336 Hz	1477 Hz
679 Hz	1	2	3
770 Hz	4	5	6
852 Hz	7	8	9
941 Hz	*	0	#

As the information highway seeks to carry more and more voice, data and video information, the frequency spectrum gets packed out, hence the present trend to use higher and higher frequency bands, although there are more problems at these higher frequencies due to propagation effects. Rain, for instance, will dramatically reduce signal strength at these higher frequencies. Newer antennas are required for higher frequencies.

The size of the antenna will depend on the frequency at which it is required to transmit or receive. Antennas radiate and receive electromagnetic signals, but they have some similarity to the ears of mammals. Consider sound waves heard by mammals. It has been observed that mammals with a longer ear span, in general, do not hear higher frequencies than those with a lower ear span. The hearing of man is within the frequency range of 20–16,000 Hz; the frequency separation to capture a change in frequency is about 3 or 4 Hz (i.e. about 0.3%). Many animals can hear below 20 Hz; e.g. sensing impending earthquakes where sound waves below 20 Hz precede earthquakes. Birds are able to navigate by winds whistling through mountains. Listed below are the maximum frequencies heard by some mammals and insects. Note that with the exception of whales and mosquitoes, a shorter ear span means an ability to hear high frequency sound signals. This indicates that antennas for higher frequency of operation will be smaller than the low frequency antennas.

Elephants	< 12 kHz
Dogs	< 40 kHz
Rats	< 70 kHz
Whales/dolphins	< 100 kHz
Moths	< 240 kHz (they can hear bat sonar)
Mosquitoes	< 150–550 Hz only

1.3   Basic characteristics and definitions of terms

Wavelength. Wavelength is the distance traveled by one cycle of a radiated electric signal. The frequency of the signal is the number of cycles per second. It follows that frequency f is inversely proportional to the wavelength. Both wavelength and frequency are related to the speed of light c. The formula is given below:

$$c = f \times \lambda,$$
$$c = 3 \times 10^8 \text{ m/s (speed of light)},$$
$$f = \text{frequency (Hz)},$$
$$\lambda = \text{wavelength (m)}.$$

Sinusoidal waveform. Why do we mostly use sinusoidal signals? There are two simple reasons for this. First, consider the differentiation of a sine term.

$$\frac{d(\sin \omega t)}{dt} = \omega \cos \omega t.$$

Although we get a cosine term, the shape of the resultant waveform is the same as that of the sine waveform. Communication electronic circuits are full of capacitors and inductors, which integrate or differentiate signals. This means that the waveform will not change if we use sinusoidal signals. Furthermore, we shall see that when high frequency signals are launched into free space, the signals propagate by means of the electric field intensity (E, V/m) and magnetic flux density (B, webers/m^2) maintaining each other through a process of spatial- and time-differentiation as defined by Maxwell's equations. Here again, using sinusoidal signals ensures that the basic waveforms of the signals carried by the electromagnetic wave do not change. If we had used a triangular waveform, for example, the resultant waveform after differentiation would be rectangular!

There is a second reason why a pure sinusoidal signal is attractive for communications: reduction of bandwidth. A single pure sinusoidal wave will only occupy a single frequency (spectrum) line in the frequency spectrum. Consider now a rectangular wave, say switching between 5 V (bit "1" in digital terms) and 0 V (bit "0"); if its Fourier transform is taken the frequency spectrum will contain a collection of sinusoidal waves all at different frequencies. This means a rectangular pulse will occupy more space in the frequency spectrum, which is undesirable since the electromagnetic frequency spectrum is rapidly becoming filled up with many communications and remote sensing channels and systems. This is also one reason why the digital communication system is more attractive than the analog frequency modulation (FM) system. In the digital system, bits 1 and 0 may be modulated onto a single frequency carrier with a phase shift to differentiate bit 1 from bit 0. In an FM system, to carry a television channel for instance, 6 MHz bandwidth is required to transmit the picture and another 100 kHz or so for the voice.

Most of the antenna analysis presented in this book will assume a single-frequency electromagnetic signal. Of course all the results we obtain apply to a multiple-frequency signal as well (e.g. an FM signal), if we analyze the signal frequency by frequency (in its frequency domain), and then sum them all up to get back the resultant signal.

Radiation. Radiation is the emission of coherent modulated electromagnetic waves in free space from a single or a group of radiating antenna elements. In order to get directive radiation beams (patterns), a group of radiating antenna elements is used to maximize signal radiation in a given direction.

Array antenna. An array antenna consists of a group of radiating or receiving antenna elements.

Beamforming. In beamforming we seek to determine the geometry of an antenna or the phase and current amplitudes (weights) for an array antenna to obtain a prescribed antenna beam (radiation pattern). This is also known as antenna synthesis.

Adaptive antenna. An adaptive antenna is an array of antennas that is able to change its radiation pattern to minimize the effects of noise, interference and multipaths and to maximize signal reception in a given direction. It may also be used to transmit signals to a mobile receiver, such that the incident signal strength at the mobile unit is always maximum. The adaptive nature of the antenna is due to the fact that it uses powerful signal processing techniques to do on-line, real-time adjustments of the radiation pattern to maximize its performance.

Switched-beam antenna. In a switched-beam antenna, a series of non-adaptive antenna beams are available at the transmitter/receiver site. By selecting a combination of beams, reception and transmission in a given direction may be maximized. This is not as dynamic as the adaptive antennas.

Smart antenna. In general, a smart antenna combines the adaptive antenna and switched-beam antenna technology. In communication systems it is essentially an adaptive antenna since it intelligently adjusts the radiation pattern to maximize the performance of the antenna. In imaging systems, it may also include processing the signals to identify targets and to perform diagnostics on the object under observation.

Decibel notation. The formula for dB calculation when dealing with voltage levels is

$$dB = 20 \log(E_1/E_2). \tag{2}$$

Example 1. Given that

$$E_1 = 900 \text{ mV/m},$$
$$E_2 = 100 \text{ mV},$$

we find

$$dB = 20 \log(900/100),$$
$$dB = 19.085.$$

The formula for dB calculation when dealing with power levels is given by

$$dB = 10 \log(P_1/P_2) \tag{3}$$

since power is proportional to (voltage)^2 or (electric field)^2.

Radiation resistance. Radiation resistance is defined in terms of transmission, using Ohm's law, as the radiated power P from an antenna divided by the square of the driving currents I at the antenna terminals. From the perspective of the transmitter circuit connected to the antenna, the power radiated by the antenna is a power loss, and is identified as the power dissipated in the radiation resistance.

Polarization. Polarization is the angle of the radiated field vector in the direction of maximum radiation. If the plane of the field is parallel to the ground, it is vertically polarized. When the receiving antenna is located in the same plane as the transmitting antenna, the received signal strength will be maximum. If the radiated signal is rotated at the operating frequency by electrical means in feeding the transmitting antenna, the radiated signal is circularly polarized. The circularly polarized signals produce equal received signal levels with either horizontal or vertical polarized receiving antennas (see Fig. 1.3).

1.4   Basic antenna parameters

All telecommunication systems use electromagnetic waves to carry information from a transmitter to the receiver. Similarly, even in digital computer systems, digital data may be transferred from the computer memory to the central processing unit by electromagnetic waves traveling along the copper conductors on the printed circuit board. In this chapter we consider the launching of electromagnetic waves into free space and capturing the signal at the receiver, using antennas. The basic concepts of some elementary antennas are described in the following sub-sections.

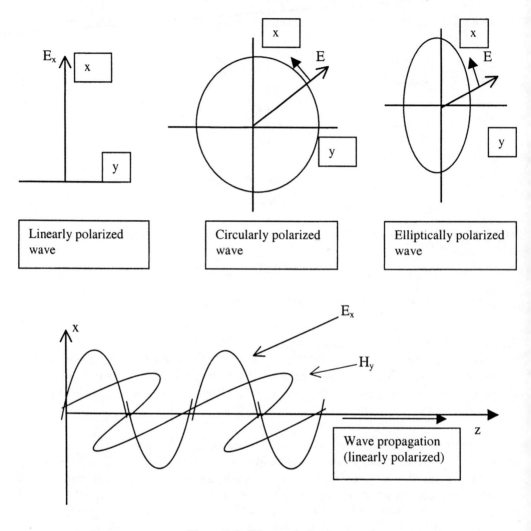

Figure 1.3: Wave polarization.

1.4.1 Antenna as a spatial filter: radiation pattern

The radiation pattern of an antenna gives an idea of the propagation of the radiated signal around the antenna. For an isotropic antenna (i.e. point radiating element), the radiation pattern will be spherical. The radiation pattern changes as the configuration of the radiating element changes. Figure 1.4 shows the radiating patterns of two directive antennas.

A major lobe (main beam) is the radiation lobe containing the maximum radiation power. Minor lobes are any lobes other than the main lobe. The minor (or side) lobes

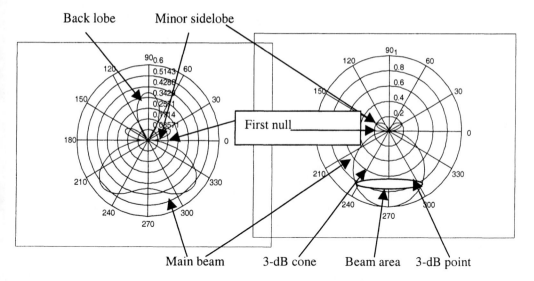

Figure 1.4: Antenna radiation patterns and beam lobes.

are unwanted since they radiate information (or electromagnetic energy) in unwanted and unintended directions. A back lobe is a minor lobe that occupies a direction opposite to that of the major lobe. Minor lobes should be minimized in order to get the antenna to radiate and receive only in the prescribed or intended direction.

Narrow or pencil beams are preferred for antennas that must be highly directive. In some applications like radio broadcasts where the radio program should be transmitted in all directions, the antenna radiation pattern should have a wide main beam that spans the entire 360° in the ground (i.e. H-) plane.

For a fictitious isotropic antenna (point radiating element) the radiation pattern will be perfectly spherical. In reality what comes closest to an isotropic antenna is a short or infinitesimal dipole antenna. The radiation pattern depends on the antenna used, and the combination of a number of antenna elements (as in array antennas). Let us say that the radiation pattern of a transmitting antenna at a distance r from the antenna is defined by $E(r,\theta,\phi) = E_m f(\theta,\phi)$, where $E(r,\theta,\phi)$ is the maximum electric field at a distance r from the antenna, and the function $f(\theta,\phi)$ defines the radiating pattern. The transmitting antenna is assumed to be placed at the origin. Although we discuss the radiation pattern of an antenna for transmitting, it is the same pattern when the antenna is receiving signals at the same frequency as when transmitting. Now E_m is the maximum of the maximum electric field measured in a circle of radius r around the antenna: $E_m = \text{Maximum}\{E(r,\theta,\phi)\}$. The time domain electric field at position (r,θ,ϕ) is given by $E(r,\theta,\phi) \cos(\omega t - k r)$. Therefore the power density at (r,θ,ϕ) is given by $E(r,\theta,\phi) = P_m f^2(\theta,\phi)$, where $P_m = E_m^2/2\eta$ W/m^2.

The fact that the radiation pattern has nulls in certain (θ,ϕ) directions and peaks in other directions implies that the antenna acts like a filter in the spatial (r,θ,ϕ) domain. Hence an antenna may be considered as a spatial filter, a concept that will be used in antenna beamforming or antenna synthesis. We shall see that an antenna, in addition to being a spatial filter, is also a temporal filter with its response peaking at a certain resonant frequency and not allowing signals at other frequencies to be transmitted or received.

1.4.2   Antenna gain and beam width

Gain is the signal level produced (or radiated) by an antenna relative to that of a standard reference dipole antenna. It is used frequently as the figure of merit. Gain is closely related to directivity, which in turn is dependent upon the radiation pattern. High values of gain are usually obtained with a reduction in beam width. Gain can be calculated only for simple antenna configurations. Consequently, it is determined by measuring the performance relative to a reference dipole. The reference dipole is either the omni-directional isotropic antenna or the half-wave dipole antenna.

Hence the directivity, gain and efficiency of antennas are scalar quantities that are used to express the antenna radiation characteristics. All these are defined for a particular antenna with respect to a reference isotropic antenna. If the total power radiated is P_r W then the density of power radiated by an equivalent isotropic antenna at distance r is $P_r/4\pi r^2$ W/m^2. The directivity (D) of an antenna may alternatively be expressed as

$$D = \frac{P_m}{P_r/4\pi r^2} = \frac{r^2 P_m}{P_r/4\pi} = \frac{S_{max}}{P_r/4\pi}, \tag{4}$$

where P_m = maximum power density radiated by the antenna in W/m^2; P_r = total power radiated by the antenna in W; $S_{max} = r^2 P_m$ maximum value of radiation intensity in W/unit solid angle; $P_r/4\pi$ = radiation intensity of an isotropic antenna. The directivity of the antenna will be entirely determined by the radiation pattern of the antenna. A narrower beam with fewer sidelobes will mean a highly directive antenna. The gain (G) of an antenna is given by

$$G = \frac{S_{max}}{P_T/4\pi} \tag{5}$$

$$< D \quad \text{for } P_r < P_T. \tag{6}$$

The gain of an antenna is usually less than the directivity of the antenna due to losses in the antenna. Sometimes the gain is specified in decibels: $G_{dB} = 10 \log_{10}(G)$ dBi. The efficiency of an antenna is given by

$$\eta_e = \frac{P_r}{P_T} = \frac{G}{D}. \tag{7}$$

When there is no loss in the antenna, i.e. $P_r = P_T$, the efficiency of the antenna is 100%. The gain G, sometimes called the isotropic gain, of an antenna does not mean that the antenna amplifies the overall power input. Indeed, if ohmic losses in the antenna are neglected, then the power delivered to the antenna is equal to the power radiated out, provided that there is no reflected energy at the antenna due to impedance mismatch.

The gain of an antenna denotes the increase in maximum power density radiated by a given antenna A_1 compared to the power density radiated by an isotropic antenna A_i. An isotropic antenna A_i radiates equal power density over the entire surface of a sphere surrounding it, with the antenna at the center of the sphere. Thus at a distance r from the antenna the power density is $P_r/4\pi r^2$ W/m^2, where P_r is the total power radiated (in W) by the isotropic antenna. When the same amount of power P_r is radiated by another antenna A_1, the maximum power density it radiates will be $GP_r/4\pi r^2$ W/m^2, where G is the gain of antenna A_1. Some manufacturers define the gain of an antenna with reference to a half-wave dipole antenna instead of an isotropic antenna. In this case, the area $4\pi r^2$ over which an isotropic antenna radiates has to be replaced by the smaller area over which a dipole antenna radiates, which means that the dipole gain G_d of a given antenna will be smaller than its isotropic gain G_i, i.e. G.

The total efficiency of an antenna η_e is defined by taking into account the power losses at the antenna terminals and within the antenna structure itself. Power losses may be due to both reflections at the antenna terminals because of poor impedance matching between the antenna and the transmission line, and the $I^2 R_o$ losses due to the ohmic resistance R_o of the antenna conductor and dielectric materials. Hence

$$\eta_e = \eta_r \eta_c \eta_d, \tag{8}$$

where η_r = reflection efficiency, η_c = conduction efficiency and η_d = dielectric efficiency.

Half-power beam width (HPBW) θ_B in the plane of the antenna is the angular width of the radiation pattern, where the power level of the received signal is down by 50% (3 dB) from the maximum power density. Let Ω_B be the HPBW solid angle (in steradians) measured between the 3 dB points on the radiation pattern. We assume that all radiated power P_r is confined to the half-power beam; i.e. we ignore the radiation outside the HPBW angle (θ_B radians) as being small and negligible. For most antennas the HPBW is approximately equal to FNBW/2, where FNBW is the first null beam width. The beam area is given by

$$A_b = r^2 \Omega_B = r^2 \frac{\pi}{4} \theta_B^2, \tag{9}$$

where

$$\Omega_B = \frac{\pi}{4} \theta_B^2. \tag{10}$$

If the total radiated power is P_r, then the maximum power density, assuming that the power density is uniform and constant over the entire beam area A_b,

$$P_m = P_r / A_b = P_r/r^2 \, \Omega_B.$$

Therefore the directivity of the antenna is given by

$$D = \frac{P_{max}}{P_r / 4\pi r^2}$$

$$= 4\pi / \Omega_B. \tag{11}$$

For a lossless antenna $P_T = P_r$, and thus

$$G = D = 4\pi/\Omega_B,\tag{12}$$

where Ω_B is the HPBW solid angle (in steradians) measured between the 3 dB points. This is an approximate formula and applies to all antennas. For an exact calculation of D and G we must not assume that the power density over the beam area is constant and uniform. Since maximum power density $P_m = (E_m^2/2\eta)$, the maximum electric field radiated to a distance r from the antenna is given by $E_m = (1/r)(DP_r\,\eta/2\pi)^{1/2}$ V/m.

The FNBW is defined as the angular width between the two null points on either side of the main beam of the radiation pattern. The FNBW is always greater than the HPBW.

Example 2. An antenna has a gain $G = 61$ dB. Find θ_B.

$$G = 10^{61/10} = 125.89 \times 10^4,$$

$$\Omega_B = \frac{4\pi}{G} = 0.0998 \times 10^{-4},$$

$$\theta_B = \sqrt{\frac{4}{\pi}}\sqrt{\Omega_B} = 0.0036\ \text{rad} = 0.204°.$$

Example 3. A satellite station in synchronous orbit has an antenna with a HPBW $\theta_B = 0.2°$. At synchronous orbit, $d = 36{,}000$ km is the distance between the satellite and the earth. Find the area of the earth covered by the satellite antenna beam.

$$\theta_B = 0.2° = 3.49 \times 10^{-3}\ \text{rad},$$

$$\Omega_B = \frac{\pi}{4}(3.49 \times 10^{-3})^2,$$

$$\text{area of spot} = r^2\Omega_B = 12.4 \times 10^9\ \text{m}^2.$$

For gain $G(\theta,\phi)$, or directivity $D(\theta,\phi)$, in a given direction (θ,ϕ) other than the direction of maximum radiation, the P_m of eqns (4) and (5) must be replaced by $P_i(\theta,\phi)$, where $P_i(\theta,\phi)$ is the power density radiated in the direction (θ,ϕ). $P_i(\theta,\phi) = E(\theta,\phi)^2/(2\eta)$.

1.4.3   Effective aperture

The concept of an effective aperture captures the idea of an antenna being open to capture power radiated by another (transmitting) antenna, which falls within its aperture. If P_i is the incident power density (W/m^2) at the antenna and P_R (W) is the total power received or captured by the antenna, the effective aperture is defined by

$$A_{em} = \frac{P_R}{P_i} = \frac{\frac{1}{2}I^2R_T}{P_i}.\tag{13}$$

For maximum power transfer

$$R_A = R_r + R_o = R_T,$$
(14)

$$X_A = -X_T,$$
(15)

where R_A = antenna resistance, R_r = antenna radiation resistance, R_o = antenna ohmic resistance, R_T = termination resistance, X_A = antenna reactance and X_T = termination reactance. Hence the current in the receiver circuit of the antenna is given by

$$I_T = V_T/(Z_A+Z_T)=V_T/2R_T$$
$$=V_T/[2(R_r + R_o)].$$
(16)

Hence the effective aperture area of the antenna is given by

$$A_{em} = \frac{V_T^2}{8P_i(R_r + R_L)} = \frac{V_T^2}{8P_iR_r} \quad \text{for } R_L \ll R_r.$$
(17)

For a short dipole of length L and an incident electric field E, using the following parameters ($R_r = 80(\pi L/\lambda)^2$, the radiation resistance; $D = 1.5$, the directivity; $P_i = E^2/2\eta_0 = E^2/2(377)$ W/m^2, the incident power density at the antenna; $V_T = EL$, the voltage induced along the antenna wire), we get

$$A_{em} = 0.12\lambda^2 \text{ m}^2$$
$$= \frac{\lambda^2}{4\pi}D$$
(18)

for a short or infinitesimal dipole.

A useful equation for the gain of an antenna in terms of its maximum effective aperture area A_{em}, which we shall define in Chapter 2, is $G =(4\pi A_{em})/\lambda^2$. Hence $G \propto A_{em}$ and, further, $G \propto f^2$. Note that in general the gain of an antenna increases as the operating frequency f is increased; $f = c/\lambda$, where c is the velocity of light (approximately equal to 3×10^8 m/s) and λ is the wavelength. For an elemental dipole, $G = D = 1.5$, for a half-wave dipole $G = D = 1.64$.

1.4.4 Operation zones

Three zones are defined for an antenna. These zones depend on the size of the antenna D (e.g. D = length of antenna for a wire type antenna) and the wavelength λ.

The near-field zone is the distance r from the antenna that satisfies the following relation:

$$r < r_1 = 0.62D\sqrt{\frac{D}{\lambda}}.$$
(19)

In the near-field region of an antenna is the region immediately surrounding the antenna where only the reactive field dominates.

The radiating near-field (Fresnel) zone is defined by distance r which lies in the range

$$r_1 < r < r_2 = 2D^2 / \lambda. \tag{20}$$

In the radiating near-field or intermediate field region the radiation field dominates and the angular field distribution is dependent on the distance from the surface of the antenna. For short or infinitesimal antennas where the size of the antenna is very small compared to the wavelength, this region does not exist.

In the far-field (Fraunhofer) zone, in which most antennas operate, the receiving antenna is far away from the transmitting antenna:

$$r > r_2. \tag{21}$$

In the far-field region of an antenna the angular field distribution may be assumed to be independent of the distance from the antenna. Although most communication antennas operate in the far-field zone, in biomedical imaging the antennas (or coils) operate at frequencies of the order of 160 MHz and are only about 0.5 m away from the radiating source. In mobile communications, it is possible for the receiver antenna to pick up near fields when it is close to the base station, as in the case of indoor picocells.

1.4.5   Antenna as a temporal filter: bandwidth

The bandwidth of an antenna may be defined as the frequency range over which its performance is satisfactory. Depending on the system for which an antenna is designed, it is important to match the beam width of the systems and the antenna bandwidth. For instance, in mobile communications, bandwidths of 25 kHz (the AMPS system of USA) and 200 kHz (the GSM system of Europe) are common. The center frequency may be about 900 MHz. In this case, it is necessary to select an antenna which will operate at a center (resonant) frequency of say 900 MHz, and provide a bandwidth of 25 kHz (AMPS) or 200 kHz (GSM). The frequency bandwidth is a general classification of the frequency band over which the antenna is effective. Wireless communications using code-division-multiple-access (CDMA) at frequencies of 1900 and 2400 MHz demand the deployment of smart antennas with narrow bandwidths, to enable range and capacity improvement through spatial filtering as well as temporal filtering. Data rates, which partly determine the bandwidth, will increase from about 1 Mega-bits/second (Mbps) to about 5 Mbps.

(i)   Broadband antennas perform over a center frequency with a bandwidth ratio of 10:1, which is the ratio of the maximum-to-minimum frequencies of acceptable operation. A 10:1 bandwidth indicates that the maximum frequency for acceptable operation is 10 times greater than the minimum frequency. Antennas operating at 5:1 bandwidth are generally suitable for most applications, including mobile communications, which require a bandwidth of about 70 MHz.

(ii)   If the bandwidth is about 1:0.05 the antenna is called a narrow-band antenna. In a narrow-band antenna the antenna conductor diameter to length ratio must be small. If the diameter of the antenna is increased, then it becomes a wide-band antenna. Narrow-band antennas tend to detune under ice and wind loading.

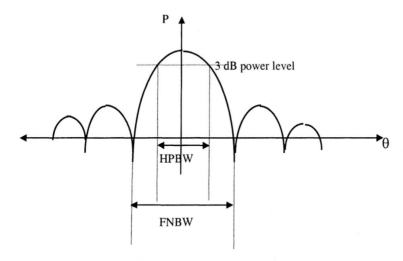

Figure 1.5: Antenna bandwidth.

Figure 1.5 shows the antenna characteristics considered as a spatial filter. The antenna transmits effectively within the HPBW angle, and beyond the FNBW we get the undesirable radiation through sidelobes. Figure 1.6 shows the characterization of an antenna as a temporal filter. Since an antenna can be considered as an RLC circuit, it has a certain frequency bandwidth. The bandwidth of an antenna may be conceptually understood by considering it as an RLC circuit. The capacitance forms between, say, the top wire and the bottom wire of a half-wave dipole antenna. The wires will have both resistance and inductance.

Now any series RLC circuit will have a resonant frequency at which the current peaks. The resonant frequency in terms of the inductance and capacitance of the antenna is given by

$$f_0 = \frac{1}{2\pi\sqrt{L_A C_A}} .$$ (22)

This is the best frequency to operate the particular antenna. It is known that the response of the RLC circuit drops off from the resonant frequency. If we operate the antenna beyond the 3 dB points, then very little of the signal will get into the antenna and even less will be radiated out. Hence we have the concept of bandwidth, which specifies the permissible frequency fluctuations of a given antenna. Given the resonant frequency and the bandwidth we will know the frequencies at which the antenna will

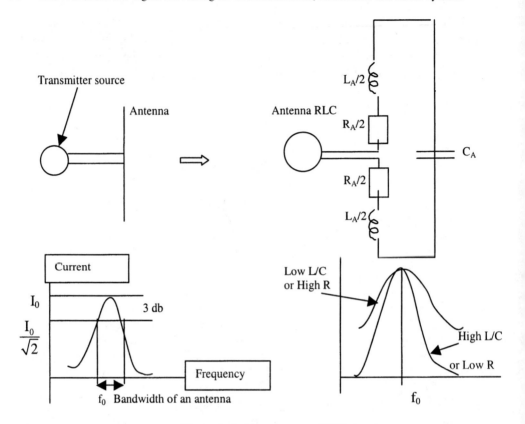

Figure 1.6: Antenna as an RLC circuit.

give satisfactory performance. For a given communication or remote sensing system, the bandwidth of the antenna must be matched with the bandwidth of the signals.

1.4.6   Antenna temperature

An antenna receives radiation from a variety of objects other than the transmitted signal, and its temperature is usually above absolute zero. $T_b > 0$ K (-273 K).

These temperatures have adverse effects on communication systems. These additional signals associated with the antenna temperature appear as noise at the receiver electronics. However in remote sensing systems, the radiation due to the finite temperature of objects other than the antenna can be captured and used to determine the material properties of the object. This is called passive remote sensing. Passive remote sensing is used in astronomy to study the properties of stars and far-off galaxies, as well as in studying the geography of land through a passive sensing antenna attached to the fuselage of an aircraft or to a satellite.

In communication systems, the physical temperature of both the antenna and other hot objects produces electromagnetic noise which interferes with the signal received. Noise power is given by $p = kT_b$ W/Hz. Hence for a bandwidth of Δf Hz, the total noise power is given by $P = p\Delta f$ W.

Each one of the following objects is at a finite temperature and will radiate an electromagnetic signal over a broad spectrum.

Sky: 3 K;   Mars: 164 K;   Earth: 290 K;   Man: 310 K.

Thus the *noise temperature* at an antenna associated with noise results from a summation of temperatures at the antenna due to many hot objects, near and far away, and is given by the total noise temperature:

$$T_A = T_{B1} + T_{B2} + \cdots .$$ (23)

Transmission line theory applied to conduction of noise along the line or waveguide, which connects the antenna to the receiver electronics, will give

$$T_a = T_A \, e^{-2\alpha d} + T_0(1 - e^{-2\alpha d}),$$ (24)

where $T_0 = 273$ K, physical temperature of the system (room temperature), and α is the attenuation coefficient of the electrical circuit.

Hence the total noise power at the receiver electronics is the sum of noise coming from the antenna (T_a) and the noise produced by the hot electronic circuitry (at a temperature T_R) itself. The total noise power at the receiver electronics operating with a bandwidth Δf Hz is given by

$$P_N = k(T_a + T_R)\Delta f.$$ (25)

For very high frequency antennas, with frequency $f > 10$ GHz,   the temperature is given by $T = hf/k$, where Planck's constant $h = 1.055 \times 10^{-34}$ J s and Boltzmann's constant $k = 1.38 \times 10^{-23}$ J/K.

1.4.7   Antenna input impedance

The antenna input impedance is the terminating resistance into which a receiving antenna will deliver maximum power. As with the gain of an antenna, the input impedance of an antenna can be calculated only for very simple formats, and instead is determined by actual measurement (see Fig. 1.7).

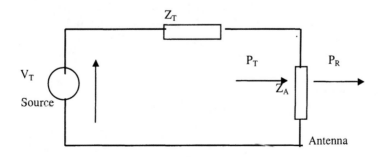

Figure 1.7: Power flow in an antenna.

electronics will be the same if the transmitter is now connected to A_y and receiver to A_x. This also implies that the transmit and receive radiation patterns of an antenna are identical.

Thus for example, if antenna A_x transmits maximum signal strength along the z-axis ($\theta = 90°$, $\phi = 90°$) and zero radiation along $\theta = 30°$, $\phi = 90°$, then if A_x is used to receive signals, it will best receive signals coming towards it along the z-axis and it will not pick up any signal coming from the $\theta = 30°$, $\phi = 90°$ direction.

Let the gain of the transmitting antenna in the direction of the receiver antenna be G_x and that of the receiver antenna in the direction of the transmitter antenna be G_y; and P_T is the total power input to the transmitting antenna. Ignoring the losses in the antennas ($G = D$) the total power radiated is P_T W. The power density at the receiver antenna is $G_x P_T / 4\pi r^2$ W/m^2. The effective aperture of the receiving antenna in the direction of the transmitting antenna is $A_{em} = D_y \cdot \lambda^2 / 4\pi = G_y \cdot \lambda^2 / 4\pi$. Hence the total power received by the second antenna at a distance r from the transmitting is given by

$$P_R = (G_x\, P_T\, /4\pi r^2) \cdot A_{em} = (G_x\, P_T\, /4\pi r^2) \cdot (G_y\, \lambda^2 / 4\pi)\ \text{W} \qquad (29)$$

or

$$P_R / P_T = G_x\, G_y\, \lambda^2 / (4\pi r)^2 . \qquad (30)$$

This is called the Friis transmission formula, and it relates the power P_R that appears at the radar receiver amplifier to the total power P_T transmitted by the radar. We have assumed that both antennas are lossless, that the impedances are all properly matched and that both antennas are correctly polarized.

The gains G_x and G_y are, in general, given by $G_x = G_{x0} f_x^2(\theta, \phi)$ and $G_y = G_{y0} f_y^2(\theta, \phi)$, where G_{x0} and G_{y0} are the maximum gains of the two antennas (e.g. $G_{x0} = G_{y0} = 1.5$ if both antennas are elemental, short dipole or $G_{x0} = G_{y0} = 1.64$ if both antennas are half-wave dipoles). The functions $f_x(\theta, \phi)$ and $f_y(\theta, \phi)$ are the normalized electric (and magnetic) field radiation patterns of the two antennas in the spherical coordinate direction (θ, ϕ). The maximum value of the radiation pattern functions is 1.0.

In mobile or wireless communication systems, the antennas are considered as short dipoles that may be approximated to isotropic (omni-directional) antennas. Hence $G = D = 1$. This means that the received power is

$$P_R = P_T\, (\lambda / 4\pi r)^2\ \text{W} \qquad (31)$$

or

$$P_{RdB}(r_0) = 10\, \log_{10}(P_T\, (\lambda / 4\pi r)^2)\ \text{dB}. \qquad (32)$$

In this equation for received power we have not considered the signal loss over the path traveled by the signal and shadowing effects. Path loss is due to losses in the finite conductivity ground over which the signal travels. Shadowing is due to objects like trees and buildings that tend to obstruct part of the signal. Thus the received power will be less than the ideal received power P_R. An expression for received power with respect to the power received at the boundary of a cell in cellular communication systems is the following:

signal power to detect the presence of a target. In stealth technology, to evade detection, missiles and aircraft are designed to geometrically and materialwise reduce the RCS.

1.6   Types of antennas

1.6.1   Elemental current antennas

Elemental current antennas are generally conducting wires that are arranged such that the signal power radiated is maximized. An inexpensive, but effective, type of wire antenna is the half-wave dipole antenna, shown in Fig. 1.8(a). The input impedance of a half-wave dipole antenna is $73 + j42.5\ \Omega$, and becomes $70 + j0\ \Omega$ if the length is slightly reduced. Hence its input impedance almost exactly matches the $70\ \Omega$ characteristic impedance of a coaxial cable. The half-wave dipole antenna is formed by bending out length $\lambda/4$ from each end of a two-wire transmission line, where λ is the wavelength of the signal to be transmitted (or received). The total length of the antenna is $\lambda/2$, as shown in Fig. 1.8(a).

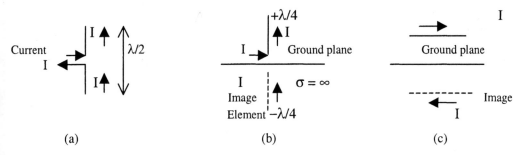

(a)   (b)   (c)

Figure 1.8: Elemental wire antennas.

An alternative way of forming a half-wave dipole antenna is to use a quarter-wavelength ($\lambda/4$) long wire placed over a perfectly conducting plane (Fig. 1.8(b)). This quarter-wavelength monopole, resulting in a half-wave dipole, is based upon the theory of images. The theory of images states that when a positive electric charge is placed above a perfectly conducting plane, the plane can be replaced by a negative electric charge placed at the mirror image point of the positive source charge. Hence an electric current moving vertically upwards above a perfectly conducting plane will have associated with it a mirror image current that is also moving vertically upwards. The input impedance of such a quarter-wavelength monopole ($36\ \Omega$) is half that of the half-wavelength dipole ($73\ \Omega$), but its directivity is twice that of the half-wave dipole, since its HPBW above ground ($39°$) is half that of the half-wave dipole ($78°$).

For over-the-horizon transmission of signals, as in the case of over-the-horizon radar, based on the reflection of electromagnetic signals from the ionosphere, horizontal dipoles such as those shown in Fig. 1.8(c) are used. The ionosphere is at a height of about 300 km above the earth. In the case of horizontally placed current carrying wire, the mirror image current flows in the opposite direction to that of the source current placed above the ground. These horizontally held antennas are also used by hobbyists to communicate to someone who is continents away on the earth. The electromagnetic communication signal travels all the way round the globe by repeatedly bouncing off the ionosphere and the earth.

The elemental type of antennas is designed to operate at frequencies ranging from 10 kHz to 1 GHz. The size of the antenna (length L) is related to the wavelength by $L/\lambda = 0.01–1$.

Similar to the half-wave dipole antenna is the loop antenna, which forms a magnetic dipole. A conducting wire is bent into a circular loop. When a high frequency source is connected across a small cut in the loop, the circular loop will radiate out the signal from the source. The radiation fields at (r,θ,φ) and radiation resistance of a loop antenna are given by

$$H_\theta = \left(\frac{ka}{2}\right)^2 I\frac{e^{-jkr}}{r}\sin\theta,\tag{38}$$

$$E_\phi = -\eta H_\theta,\tag{39}$$

$$R_r = 20\pi^2\left(\frac{2\pi a}{\lambda}\right)^2 \Omega,\tag{40}$$

where a is the radius of the loop and k $(=2\pi/\lambda = \omega/c)$ is the wave number.

1.6.2 Traveling wave antennas

Traveling wave antennas are generally long conductors or helical loops or conductors formed in a zigzag fashion over which the electric current–voltage wave travels close to the velocity of light. As the waves travel along the long conductors, they also radiate signals out into free space, thus forming an antenna. These antennas are used in the frequency range of 1–10 MHz and the size of the antennas (L) is typically of the order of $L/\lambda = 1–10$.

1.6.3 Array antennas

Array antennas are formed by a series of wire or aperture antennas, generally arranged along a single line (linear array) or in a rectangular grid (planar array). In other words, an array antenna is a collection of single antenna elements. One popular type of array antenna is the Yagi antenna used for television signal reception, shown in Fig. 1.9. The advantage of having a series of antenna elements radiating or receiving simultaneously is two-fold: increased radiation power and better ability to direct the antenna radiation pattern (beam) in a specific direction. A folded dipole antenna, instead of a half-wave dipole antenna, is used to double the amount of power received; however the input impedance of the folded dipole is about 388 Ω.

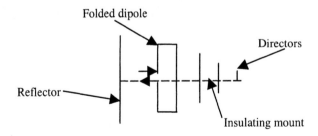

Figure 1.9: Yagi–Uda array antenna.

Such antennas are used in the frequency range of 5 MHz to 50 GHz, and the typical dimension of the antenna is $L/\lambda = 1$–100.

1.6.4   Aperture antennas

Aperture antennas are used at very high frequencies, and are formed by having apertures or windows through which the electromagnetic signals may be radiated. A widely used type of aperture antenna is the rectangular horn antenna, where electromagnetic signals traveling through a waveguide are radiated out by opening up the waveguide into a horn shape (Fig. 1.10(a)). These may be operated in the reverse direction to receive signals in a highly directive manner. The large parabolic reflectors used in satellite communication antennas function like apertures. Normally a wire antenna or a horn antenna is placed in front of the parabolic dish to collect the signals focused by the parabolic aperture (receiving) or to project the signals from an oscillator on to the parabolic dish (transmitting). The parabolic reflector arrangement is shown in Fig. 1.10(b), where a horn antenna is placed at the focal point of the parabolic reflector.

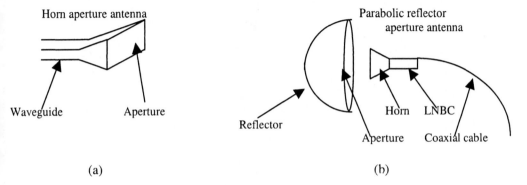

(a)   (b)

Figure 1.10: Aperture antennas.

The low noise block converter (LNBC) provides low noise amplification and frequency down conversion (e.g. from 8 GHz down-converted to 800 MHz). The down-converted signal may be carried over coaxial cable without high ohmic losses. Apertures tend to be highly directive and enable better focusing. In radio astronomy, an array of parabolic reflector type antennas is used to observe signals radiated from a very small region in the galaxy.

1.7   Waves along conductors and in free space

The launching or radiation of an electromagnetic wave from a wire antenna is shown in Fig. 1.11. A two-wire transmission line carrying high frequency signals (waves) is connected to the two wires opened out vertically. The two vertical lines here form the dipole antenna. The voltages are shown on the transmission line (going from a positive to a negatively charged line) and the electric field intensity is shown on the antenna (going from positive to negative electric charges). The voltage waves arrive from the signal source in packets of pulses, reversing direction every half wavelength. The velocity of the voltage (and current) waves along the transmission line is close to the velocity of light.

Phase I. When the top positive and bottom negative voltage pulse arrives at the antenna, there is a build-up of positive electric charges on the top wire of the antenna and negative electric charges on the bottom wire.

Phase II. As the next voltage pulse arrives, this time with negative electric charges on top, these negative electric charges gradually push out the previously present positive electric charges and form electric field lines in the reverse direction on the antenna, since the bottom antenna wire is now positively charged and the top is negatively charged. During this time, the previous collection of electric field lines going from the top wire to the bottom wire all get pushed out and crowded on the edge of the wires.

Phase III. When the next voltage pulse arrives (top positive and bottom negative), the electric field lines of phase I and phase II (which are pointing in opposite directions) join up and form one whole loop in the free space. Now these closed electric field lines, of half wavelength width, will be pushed out into space and travel out as radiation fields. This can be roughly pictured as a bubble (electric field line packets) blown out of a tube (the antenna). With each electric field line there will also be present a magnetic flux line, as stipulated by Maxwell's equations, where magnetic and electric fields generate each other at high frequencies. A whole sequence of such energy (field) packets form, each packet containing information coded from the source connected to the transmission line. Now they travel out at the velocity of light.

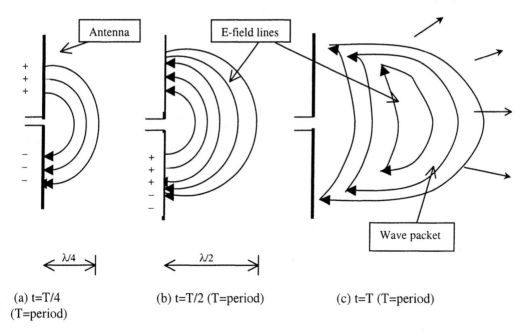

(a) t=T/4
(T=period)

(b) t=T/2 (T=period)

(c) t=T (T=period)

Figure 1.11: Formation and launching of electromagnetic wave packets.

1.8   Maxwell's equations and electromagnetic waves

1.8.1   Introduction

The basic theory that governs antenna performance and antenna signal processing is the electromagnetic theory defined by the four Maxwell equations. The set of four Maxwell's equations and the Lorentz force law are five elegant equations of natural law that underlie the entire discipline of electrical and communication engineering. Whether it is a low frequency device like an induction motor (60 Hz) or a high frequency (e.g. 10 GHz) device like the traveling wave tube (TWT) amplifier used at microwave satellites, whether it is a low frequency (50 or 60 Hz) system like the electric power system or a high frequency (e.g. 2 GHz) system like the wireless communication system, all these devices and systems are characterized and controlled by the Maxwell equations. In order to get exact equations that define the operating characteristics of a passive device like the antenna or an active device like the transistor, we must solve the Maxwell equations specifically applied to these devices. In this book we shall be looking at the solution of Maxwell equations for a variety of antennas, and the use of these solutions for antenna design and signal processing. Maxwell equations may be applied to all macroscopic electromagnetic devices and systems. The four Maxwell equations are

$$\nabla \cdot \mathbf{D} = \rho, \tag{41}$$

$$\nabla \cdot \mathbf{B} = 0, \tag{42}$$

$$\nabla \times \mathbf{H} = \mathbf{J} + \frac{\partial \mathbf{D}}{\partial t}, \tag{43}$$

$$\nabla \times \mathbf{E} = -\frac{\partial \mathbf{B}}{\partial t}, \tag{44}$$

where $\mathbf{E}$ = vector electric field intensity in V/m, $\mathbf{B}$ = vector magnetic flux density in webers/m^2, $\mathbf{D}$ = vector electric flux density in coulombs, $\mathbf{H}$ = vector magnetic field intensity in A/m, $\mathbf{J}$ = vector electric conduction current density in A/m^2 and ρ = electric charge density in coulombs/m^3. The Maxwell equations are linear since they contain E and B and their derivatives to the first power only.

The Lorentz force equation is

$$\mathbf{F} = q(\mathbf{E} + \mathbf{v} \times \mathbf{B}), \tag{45}$$

where q = electric charge in coulombs and $\mathbf{v}$ = the velocity of the electric charge in m/s.

The Lorentz force defines the physical effects of the unseen electric and magnetic fields. We observe that it is the electric field intensity E and the magnetic flux density B that are the physical parameters evidently existing; both magnetic field intensity H and electric flux density are fictitious quantities that are mathematically defined to make the field definitions complete. We have $\mathbf{H} = \mathbf{B}/\mu$ and $\mathbf{D} = \varepsilon\mathbf{E}$ and $\mathbf{J} = \sigma\mathbf{E}$, where material properties μ, ε and σ are the permeability, permittivity and electric conductivity of the material in which the fields exist. The conduction current $\mathbf{J}$ ($=\sigma\mathbf{E}$) will not exist in regions of zero conductivity, i.e. in dielectric material like free space. However, $\partial\mathbf{D}/\partial t = \partial\varepsilon\mathbf{E}/\partial t = \mathbf{J}_\mathrm{D}$, called the displacement current density, may flow in dielectric materials too. It is due to this displacement current that we can have wireless communications. Along the antenna conductor itself we have conduction currents, but once the signals are launched out from the antennas, the conduction current becomes zero, although the signal energy and information is now carried by displacement currents in free space. The displacement currents flow along the closed electric field lines of each wave packet.

1.8.2 Electromagnetic waves

Antennas are devices over which time varying electric currents flow (in wire antennas) or time varying electric and magnetic fields appear at the aperture (e.g. the open end of a waveguide antenna). These time varying currents or electric and magnetic fields at the antenna terminal will produce time varying fields just outside the antenna. In this section we shall show that these fields will constitute a propagating electromagnetic wave. With time varying fields, $\mathbf{E}$ and $\mathbf{B}$ are no longer independent as in the case of static or low frequency fields, but they are mutually coupled through the four Maxwell equations. We shall consider free space that is electric charge free (i.e. $\rho = 0$) and has zero conductivity (i.e. $J = \sigma E = 0$ in free space). Thus the four Maxwell equations (41)–(44) reduce to

$$\nabla \cdot \mathbf{E} = 0, \tag{46}$$

$$\nabla \cdot \mathbf{B} = 0, \tag{47}$$

$$\nabla \times \mathbf{B} = \frac{1}{u^2} \frac{\partial \mathbf{E}}{\partial t}, \tag{48}$$

$$\nabla \times \mathbf{E} = -\frac{\partial \mathbf{B}}{\partial t}, \tag{49}$$

where we have set the velocity of electromagnetic waves as

$$u = (\mu_0 \varepsilon_0 \mu_r \varepsilon_r)^{-1/2}. \tag{50}$$

The velocity of the wave in free space where $\mu_r = 1$, $\varepsilon_r = 1$ becomes the velocity of light in free space, $u = c = (\mu_0 \varepsilon_0)^{-1/2} = 2.9998 \times 10^8$ m/s. These four equations are not only linear, but also homogeneous since there are no electric charges or electric currents present, and every term contains either an $\mathbf{E}$ or a $\mathbf{B}$. One property of linear, homogeneous equations is that the sum of two different solutions is also a solution. Thus a general solution to these equations will consist of a sum of static as well as time varying fields. The physical origin of such static fields would be distributions of static charges and steady currents. We shall assume that such a distribution of steady electric charges and currents cannot exist in free space.

Since the divergence of both $\mathbf{E}$ and $\mathbf{B}$ is zero, the flux of the time varying $\mathbf{E}$ and $\mathbf{B}$ out of any volume is zero. Furthermore, the two curl equations couple the $\mathbf{E}$ and $\mathbf{B}$ fields together. The terms on the right hand side of the curl equations generate the curl terms on the right hand side of the equations. For a time varying $\mathbf{B}$ field, $\partial \mathbf{B}/\partial t$ is not zero; thus the curl of $\mathbf{E}$ ($\nabla \times \mathbf{E}$) is also not zero. A non-zero curl of $\mathbf{E}$ means that $\mathbf{E}$ is a function of position. This means that a time varying $\mathbf{B}$ produces a spatially varying $\mathbf{E}$. But moreover, since $\mathbf{B}$ is a function of time and the operator ∇ is independent of time, $\mathbf{E}$ must be a function of time. Hence we may conclude that a time varying $\mathbf{B}$ in space produces $\mathbf{E}$ that changes in time and position (i.e. space). Similarly the second curl equation tells us that a time changing $\mathbf{E}$ produces a $\mathbf{B}$ that changes in time and space. Thus what we have is $\mathbf{E}$ and $\mathbf{B}$ producing and supporting each other in the time and spatial domains, such that each field keeps moving (change in position or space) as each is produced by the other. We have the propagation of electromagnetic fields $\mathbf{E}$ and $\mathbf{B}$ due to each field sustaining, generating and nudging on the other.

We shall first obtain the wave equation for the electric field $\mathbf{E}$. In order to do this we need to use eqns (48) and (49) to get a single equation for $\mathbf{E}$ by eliminating $\mathbf{B}$. The curl of eqn (49) yields

$$\nabla \times (\nabla \times \mathbf{E}) = -\frac{\partial (\nabla \times \mathbf{B})}{\partial t}. \tag{51}$$

In eqn (51) the operators $\partial/\partial t$ and ∇ have been changed round since they are independent of each other. The curl eqn (48) can be used to substitute for the $\nabla \times \mathbf{B}$ term on the right hand side of eqn (51), giving us

$$\nabla \times (\nabla \times \mathbf{E}) = -\frac{1}{u^2}\frac{\partial^2 \mathbf{E}}{\partial t^2}.$$
(52)

The curl–curl term can be replaced by

$$\nabla \times (\nabla \times \mathbf{E}) = \nabla(\nabla \cdot \mathbf{E}) - \nabla \cdot \nabla \mathbf{E}.$$
(53)

Since $\nabla \cdot \mathbf{E} = 0$ for electric charge free regions, eqn (52) reduces to the wave equation

$$\nabla^2 E = \frac{1}{u^2}\frac{\partial^2 E}{\partial t^2},$$
(54)

where we have set $\nabla \cdot \nabla$ to ∇^2.

The same two curl equations in Maxwell's equations may be used to get the wave equation for the magnetic field:

$$\nabla^2 B = \frac{1}{u^2}\frac{\partial^2 B}{\partial t^2}.$$
(55)

Equations (54) and (55) are vector wave equations and u is the velocity of propagation of the waves. It should be noted that the displacement current term $(1/u^2)$ $(\partial E/\partial t)$ on the right hand side of eqn (48) is crucial to the existence of electromagnetic waves, for without this term Maxwell's equations would not yield wave equations for $\mathbf{E}$ and $\mathbf{B}$. Notice also that the two vector wave equations are not independent of each other. This is because $\mathbf{E}$ and $\mathbf{B}$ are related by the two curl eqns (46) and (47). We will employ this connection below. We should remark that the vector wave equation satisfied by $\mathbf{E}$ and $\mathbf{B}$ describes the propagation of undamped waves. This is just what we should expect as there are no charges or currents present to interact with the field and hence take energy from them.

The vector wave eqns (54) and (55) describe the propagation of waves in empty space in full generality; i.e. the wave fronts can be plane or spherical or cylindrical or any shape that we care to consider. However, from now on we will confine our attention to plane waves as these are the simplest type of wave. Plane waves are also of particular practical interest because a limited portion of a non-planar wave front that is far from its source approximates closely to a plane.

We will choose a rectangular Cartesian coordinate system because it has the appropriate symmetry, and consider waves that propagate in the positive z direction. The two vector wave equations can now be written as a set of six equations in the Cartesian components of the two fields. For the x component of the $\mathbf{E}$ field we have

$$\nabla^2 E_x = \frac{1}{u^2}\frac{\partial^2 E_x}{\partial t^2},$$
(56)

with similar equations for the y and z components of the $\mathbf{E}$ field and for the x, y and z components of the $\mathbf{B}$ field.

Over the unbounded plane wave fronts the magnitudes and directions of the fields will not vary, so the partial derivatives of the field components with respect to x and y will be zero. Thus

$$\frac{\partial}{\partial x}(E) = 0,$$
$$\frac{\partial}{\partial y}(E) = 0. \tag{57}$$

The second partial derivatives of the field components with respect to x and y will also be zero, so the three-dimensional wave eqns (56) reduce to the set of one-dimensional wave equations

$$\frac{\partial^2 E_x}{\partial z^2} = \frac{1}{u^2}\frac{\partial^2 E_x}{\partial t^2},$$

$$\frac{\partial^2 E_y}{\partial z^2} = \frac{1}{u^2}\frac{\partial^2 E_y}{\partial t^2}, \tag{58}$$

$$\frac{\partial^2 E_z}{\partial z^2} = \frac{1}{u^2}\frac{\partial^2 E_z}{\partial t^2},$$

with similar equations for the components of the **B** field. Before proceeding further, we shall assign a certain vector direction for electric field **E**, by setting

$$\mathbf{E} = \mathbf{u}_x E_x + \mathbf{u}_z E_z, \tag{59}$$

where $\mathbf{u}_x$ and $\mathbf{u}_z$ are unit vectors. Now if we take the curl of this electric field, remembering that $\partial/\partial x = \partial/\partial y = 0$, we only get one non-zero component which is $\mathbf{u}_y dE_x/dz$, which from Maxwell's eqn (49) is equal to $-\mu\partial \mathbf{H}/\partial t$. Hence $\mathbf{u}_y dE_x/dz = -\mu\partial\mathbf{H}/\partial t$. This in turn means that the magnetic field intensity can have only a y-directed component, i.e. $\mathbf{H} = \mathbf{u}_y H_y$. Now take the curl of $\mathbf{H} = \mathbf{u}_y H_y$, and we get $-\mathbf{u}_x dH_y/dz$, which from Maxwell's eqn (48) is equal to $\varepsilon\partial\mathbf{E}/\partial t$. Therefore the electric field can only have the x-directed component and $E_z = 0$. Therefore we have

$$\mathbf{E} = \mathbf{u}_x E_x. \tag{60}$$

Consider now what the first two Maxwell's eqns (48) and (49) have simplified for the case of plane wave: $\mathbf{u}_y dE_x/dz = -\mu\partial\mathbf{H}/\partial t$ and $-\mathbf{u}_x dH_y/dz = \varepsilon\partial E_x/\partial t$. Differentiating the second equation with respect to z, and substituting from the first equation for dE_x/dz, we get

$$\frac{\partial^2 H_y}{\partial z^2} = \frac{1}{u^2}\frac{\partial^2 H_y}{\partial t^2}, \tag{61}$$

which defines a magnetic wave traveling in the z direction. Similarly we get

$$\frac{\partial^2 E_x}{\partial z^2} = \frac{1}{u^2}\frac{\partial^2 E_x}{\partial t^2}. \tag{62}$$

For an electromagnetic source oscillating at a harmonic frequency of ω, setting $\partial/\partial t = j\omega$, we can show that eqns (61) and (62) have the harmonic traveling wave solutions

$$B_y = B_0 \cos(\omega t \pm kz), \tag{63}$$

$$E_x = E_0 \cos(\omega t \pm kz), \tag{64}$$

where k is the wave number, and is related to the wavelength λ by $k = 2\pi/\lambda = \omega/u$. Note that a half-wavelength long antenna designed for operation at a radian frequency of f ($=\omega/2\pi$) will have a length $u/2f$ in a medium with relative permittivity ε_r (e.g. $\varepsilon_r = 80$ for seawater), and $c/2f$ for free space. Hence the antenna designed for operation in seawater will have to be shorter by a factor of about 0.11 ($=\varepsilon_r^{-1/2}$) to that used in free space for the same frequency of operation.

The electric and magnetic fields have waves (and hence electromagnetic energy and electronic information) traveling in the $+z$ and $-z$ directions. This is normally the case with a simple wire antenna used in mobile phones; signals are radiated in both directions deduced from the idea of an omni-directional monopole antenna. If we want the waves to be directed in one direction only, say in the $+z$ direction, we must find a way of redirecting the energy traveling in the $-z$ direction to the $+z$ direction. This may be done by simply placing a reflecting plane just behind in the antenna to turn back the $-z$-directed waves to the $+z$ direction as well; by doing this all the energy put into the antenna is radiated in the $+z$ direction instead of being divided into half for transmission in both directions.

An alternative form of expressing the solution of the wave eqns (61) and (62) is

$$E_x = E^+ \exp(-kz) + E^- \exp(kz), \tag{65}$$

$$H_y = (E^+/Z) \exp(-kz) - (E^-/Z) \exp(kz), \tag{66}$$

where the wave or intrinsic impedance $Z = E^+/H^+ = E^-/H^- = (\mu/\varepsilon)^{1/2}$ Ω. Notice that the electric and magnetic fields of the wave are in phase with each other. For free space $\mu = \mu_0$, $\varepsilon = \varepsilon_0$ and thus $E^+/H^+ = \mu_0 E_0/B_0 = (\mu_0/\varepsilon_0)^{1/2}$. From this substitution, recalling that $u = c = (\mu_0\varepsilon_0)^{-1/2}$ for free space, we also find that

$$E_0 = cB_0. \tag{67}$$

The Maxwell equations are linear. Hence the principle of superposition applies and more general solutions for spherical waves can be obtained by adding together different plane-polarized electromagnetic wave solutions. An interesting example of such a solution is the case where two plane-polarized electromagnetic waves of the same amplitude, both propagating in the positive z direction, are added together. If one of the waves has its E vector plane-polarized in the x direction whilst the other has its E vector plane-polarized in the y direction, and if there is a phase difference of 90° between the waves, the resultant E vector and B vector have constant amplitude and rotate in the plane of the wave front as the wave advances. Such a wave is described as being circularly polarized. Such waves are found in microwave communication systems, where information capacity at a particular frequency may be doubled by transmitting two orthogonal electromagnetic waves.

In the foregoing discussion, we assumed that the medium through which the electromagnetic wave travels is lossless, i.e. conductivity $\sigma = 0$. If this is not the case, our analysis for the lossless case applies, except that we have to replace the permittivity ε by a complex permittivity ε^c. This comes about as follows. For lossy media, the conductivity is reflected in the following Maxwell's equations:

$$\nabla \times \mathbf{H} = \mathbf{J} + \frac{\partial \mathbf{D}}{\partial t}, \tag{68}$$

which may be rewritten as

$$\nabla \times \mathbf{H} = \sigma \mathbf{E} + \frac{\varepsilon \partial \mathbf{E}}{\partial t}, \tag{69}$$

where the first term on the right hand side is zero if the medium is lossless. Thus in the discussion so far, only the second term on the right hand side, which may be expressed as $j\omega\varepsilon E$ for a harmonic signal, existed. Now for the complete eqn (69) the right hand side may be rewritten as $(\sigma + j\omega\varepsilon)E = j\omega(\varepsilon - j\sigma/\omega)E = j\omega\varepsilon^c E$, where $\varepsilon^c = (\varepsilon - j\sigma/\omega)$. Once we make this switch from permittivity ε by a complex permittivity ε^c, the foregoing discussion for the lossless case may be used for the lossy case. The final result for waves in the lossy medium case is given by the following two equations:

$$E_x = E_0 \exp(\pm\alpha z) \cos(\omega t \pm \beta z), \tag{70}$$
$$B_y = B_0 \exp(\pm\alpha z) \cos(\omega t \pm \beta z), \tag{71}$$

where α is the loss attenuation constant and β is the phase constant. The amplitude of the wave is attenuated as it propagates through the medium; after traveling over a distance called the skin depth $z = \delta = 1/\alpha$, the wave amplitude would have been attenuated by a factor e^{-1}. Beyond this distance δ, the signal will be too weak for a receiver system to pick up. The solution of the wave equation in lossy media may also be expressed as follows:

$$E_x = E^+ \exp(-\gamma z) + E^- \exp(\gamma z), \tag{72}$$
$$H_y = (E^+/Z) \exp(-\gamma z) - (E^-/Z) \exp(\gamma z), \tag{73}$$

where propagation constant $\gamma = \alpha + j\beta = [j\omega\mu(\sigma + j\omega\varepsilon)]^{1/2}$ and intrinsic impedance $Z = [j\omega\mu/(\sigma + j\omega\varepsilon)]^{1/2}$. E^+ and E^- are two constants associated with the amplitude of the forward (positive z directed) and backward (negative z directed) waves, respectively, at $z = 0$. When the medium is a very good conductor such that $\sigma \gg \omega\varepsilon$, then $\alpha = \beta = (\omega\mu\sigma/2)^{1/2}$.

1.8.3   Energy in the electromagnetic field

Consider the following two Maxwell's equations:

$$\nabla \times \mathbf{H} = \mathbf{J} + \frac{\partial \mathbf{D}}{\partial t}, \tag{74}$$

$$\nabla \times \mathbf{E} = -\frac{\partial \mathbf{B}}{\partial t}. \tag{75}$$

Subtract the dot product of eqn (74) with $\mathbf{E}$ from the dot product of eqn (75) with $\mathbf{H}$:

$$\mathbf{H} \cdot \nabla \times \mathbf{E} - \mathbf{E} \cdot \nabla \times \mathbf{H} = -\mathbf{H} \cdot \frac{\partial \mathbf{B}}{\partial t} - \mathbf{E} \cdot \frac{\partial \mathbf{D}}{\partial t} - \mathbf{E} \cdot \mathbf{J}. \tag{76}$$

Equation (76) may be rewritten as

$$\nabla \cdot (\mathbf{E} \times \mathbf{H}) = -\mathbf{H} \cdot \frac{\partial \mathbf{B}}{\partial t} - \mathbf{E} \cdot \frac{\partial \mathbf{D}}{\partial t} - \mathbf{E} \cdot \mathbf{J}. \tag{77}$$

Taking the volume integral of eqn (77), we get

$$\iint_s (\mathbf{E} \times \mathbf{H}) \, ds = -\iiint_v \cdot \frac{\partial \mu \mathbf{H}^2}{2 \partial t} \, dv - \iiint_v \cdot \frac{\partial \varepsilon \mathbf{E}^2}{2 \partial t} \, dv - \iiint_v \mathbf{E} \cdot \mathbf{J} \, dv. \tag{78}$$

There are four terms in the energy eqn (78); these are physically interpreted as follows:

(i) The $\mathbf{P} = \mathbf{E} \times \mathbf{H}$ term on the left hand side is called the Poynting vector. It denotes the flow, or radiation of electromagnetic energy out of a surface s. The average power in an electromagnetic wave is found from $((1/2)\mathrm{Re}(\mathbf{E} \times \mathbf{H}^*))$, where the asterisk stands for conjugate.

(ii) The $-(\partial \mu \mathbf{H}^2/2 \partial t)$ term indicates the rate of decay (note the negative sign) in the magnetic energy $\mu H^2/2$ J/m^3 stored in a volume v. It is this decay of energy around an antenna that produces the radiated power $\mathbf{E} \times \mathbf{H}$ W/m^2 flowing out from the antenna. The term $\mu H^2/2$ J/m^3 is associated with the energy stored in inductive circuits or devices.

(iii) The $-(\partial \varepsilon \mathbf{E}^2/2 \partial t)$ term indicates the rate of decay of electric energy $\varepsilon E^2/2$ J/m^3 stored in a volume v. The term $\varepsilon E^2/2$ J/m^3 is what is important in the rare capacitive type of antennas. It is this term that exists in electronic transistors providing power amplification as well as in capacitors in the form of stored energy.

(iv) The fourth term $E \cdot J = \sigma E^2$ W/m^3 (since $J = \sigma E$) is associated with the ohmic power loss due to the finite conductivity of the medium through which the wave travels. Typical (conductivity in S/m, relative permittivity) values encountered in wireless communication channels are $(4, 80)$ for seawater, $(0.001, 80)$ for fresh water, $(0.02, 30)$ for swampy land, $(0.004, 13)$ for forest terrain, $(0.002, 14)$ for rocky terrain and $(0.001, 10)$ for sandy terrain. The values do not reflect the frequency dependence of the parameters.

For sinusoidal signals, we may replace $\partial/\partial t$ by $j\omega$, and thus the energy transported by the electromagnetic wave is $U = (1/2)\varepsilon E^2 + (1/2)\mu H^2$ J/m^3. Since $E = (\mu_0/\varepsilon_0)^{1/2} H = cB$, the energy carried by the electric field is equal to the energy carried by the magnetic field. We get

$$U_{elect} = \tfrac{1}{2}\varepsilon_0 E^2, \qquad U_{mag} = \tfrac{1}{2}\varepsilon_0 c^2 B^2, \tag{79}$$

and the total energy $U = \varepsilon_0 E^2 = \mu_0 H^2$. The amplitude of the Poynting vector

$$P = E^2/\eta = c\varepsilon_0 E^2 \text{ W/m}^2 \tag{80}$$

and the average power transported by the electromagnetic wave is $(1/2)\text{Re}\{E \times H^*\}=$ $(1/2)E^2/\eta =(1/2)c\varepsilon_0 E^2$ W/m^2.

It can be shown that for an electromagnetic wave propagating in a good conductor, nearly all of the wave's energy is carried by the magnetic component. In the case of an isotropic homogeneous medium, the Poynting vector becomes

$$\mathbf{P} = \mathbf{E} \times \mathbf{H} \tag{81}$$

and the energy density becomes

$$U = \tfrac{1}{2}\left(\varepsilon E^2 + \mu H^2\right). \tag{82}$$

Note that these expressions reduce to those obtained previously when $\varepsilon = \varepsilon_0$, $\mu = \mu_0$ and $H = B/\mu_0$.

The ratio of electric to magnetic energy density is therefore

$$\frac{\tfrac{1}{2}\varepsilon E_0^2}{\tfrac{1}{2}\mu H_0^2}. \tag{83}$$

Now for a good conductor

$$\sigma \gg \omega\varepsilon, \tag{84}$$

and the intrinsic impedance

$$Z = [j\omega\mu/(\sigma + j\omega\varepsilon)]^{1/2} \approx [j\omega\mu/\sigma]^{1/2}. \tag{85}$$

Since $E_0/H_0 = \eta$, from eqns (83) and (85) we get

$$\frac{\tfrac{1}{2}\varepsilon E_0^2}{\tfrac{1}{2}\mu H_0^2} = \frac{\varepsilon\omega}{\sigma}. \tag{86}$$

Now $\varepsilon\omega/\sigma$ is a very small quantity for a good conductor for all frequencies up to and including optical frequencies. Thus to a very good approximation the energy is located in the magnetic component of a wave propagating in a conductor. In contrast, we earlier found that when an electromagnetic wave propagates through a vacuum its energy is shared equally between the electric and magnetic components. Hence for media like seawater, which acts like a good conductor at most undersea communications frequencies, antennas should be designed to capture the magnetic fields more than the electric fields. In free space, most antennas like the electric dipole are designed to capture the electric field, whereas some antennas like the loop antenna are designed to capture the magnetic field.

1.9   Points to note when purchasing or designing antennas

In this section we shall describe aspects of antenna design and design parameters with reference to a particular mobile communication system. Figure 1.12 shows an aircraft in flight. It is required that we design an antenna that can be mounted on top of the aircraft to communicate with a satellite.

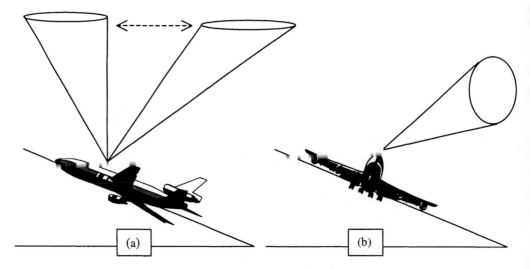

Figure 1.12: Antenna for aircraft to satellite communications.

Figure 1.12(a) shows two positions of the required azimuth scanning beam for the aircraft antenna to preserve contact with the satellite as it flies past the satellite. Figure 1.12(b) shows the elevation beam of the same antenna, required for the aircraft antenna to make contact with the satellite positioned on the left hand side of the aircraft. The following points should be observed when designing or selecting an antenna:

1. Have a thorough understanding of the system in which you will be using the antenna. Take, for instance, the challenging problem of designing an antenna for satellite to aircraft communication (Taira et al., 1991). A geostationary satellite like the ETS-V satellite orbits at an altitude of 36,000 km above the earth. An aircraft might cruise at an altitude of about 20 km. Therefore the antenna radiation pattern must be sufficiently tilted upwards so that the aircraft antenna can *look up* at the satellite; this means that the antenna beam must have an elevation beam. Further, the ETS-V satellite is stationary at 150°E, longitudinal. Consider an aircraft flying from Singapore (Changi airport) to Tokyo (Narita airport), and then from Tokyo to Anchorage. When the aircraft is flying Singapore–Tokyo–

Anchorage, the satellite is to the right hand side (RHS) of the aircraft. Hence the antenna radiation pattern should be focused to the RHS of the aircraft. Over the Singapore to Tokyo route, which is the first portion of the flight route, the satellite station is to the right front end direction of the aircraft. As the aircraft flies past the satellite, the beam should now look back at the satellite. Hence the antenna beam should have an azimuth beam width scanning azimuth angle 110° to about azimuth angle 51°. The elevation of the beam should be about 30° for the Singapore to Tokyo route. For the Tokyo to Anchorage route the elevation of the beam should be about 5°, and the azimuth beam should scan from 150° to 130°. Figure 1.13(a) shows the antenna azimuth beams for the forward flight from Singapore to Anchorage.

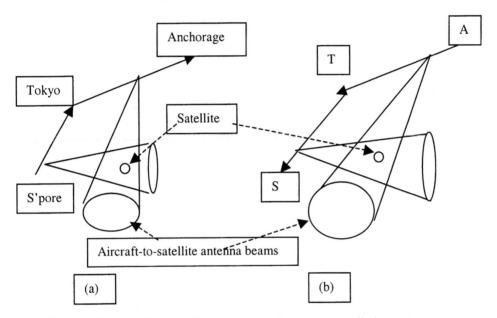

Figure 1.13: Aircraft-to-satellite communication antenna radiation patterns.

Similarly, on the way back from Anchorage to Singapore (Fig. 1.13(b)), the aircraft antenna beam should have the following left hand side (LHS) looking beams: from Anchorage to Tokyo the azimuth ranges from 300° to 330°, and the elevation is 5°; from Tokyo to Singapore the azimuth of the beam should span 290°–230°, and the elevation of the beam above the plane of the aircraft is about 36°. Obviously to design such an antenna requires not only sound antenna hardware, but also signal processing software to keep changing the direction of the radiation pattern in which the antenna has to communicate. In practice, a two-element array antenna is used to get the elevation of the beam and an eight-element array antenna is used to get the azimuth control of the beam. However, LHS and RHS communication with such tight control of beams requires two identical antennas placed on either side of the aircraft; the array antenna on the RHS of the aircraft will be used during the Singapore to Anchorage flight. The second array will be used during the Anchorage to Singapore flight. Phase dividers are used to change the phase of the signal supplied to each element of the array antenna to get a scan beam that changes with the movement of the aircraft-antenna. Such challenging antenna design problems demand optimum design.

2. Design and select the antenna to match the transmitter and receiver frequency, as well as the bandwidth of the signals. In the aircraft to satellite communication system, for instance, the operating frequency may be 1.6/1.5 GHz. Hence the antenna must be able to perform well at carrier frequencies of 1.6 and 1.5 GHz, recalling that at each carrier frequency the signal frequency will vary over the bandwidth of about 7% of the carrier frequencies. Normally, mechanically light and small patch antennas tend to have bandwidths of about 3%.

3. Decide on the mechanical restrictions to be imposed on the antenna. It will be impractical to mount an antenna weighing above 20 kg or so on an aircraft. The mechanical structure of the antenna should be able to withstand vibrations in the range of 5–2 Hz. To reduce wind resistance, in mobile applications, it may be necessary to use bandwidth-inefficient microstrip patch antennas.

4. Determine the antenna performance in changing atmospheric conditions. In a typical flight from Singapore to Anchorage the antenna will be expected to perform in temperatures varying from −60° to 75°F. The voltage standing wave ratio (VSWR) of the antenna can vary by as much as 40% for such drastic temperature variations, giving rise to the antenna being poorly matched to the transmitter/receiver electronic circuitry.

5. Where the antenna beam has to be steered, determine the effect of beamsteering on antenna gain.

6. Study the gain of the antenna. Antenna gain tends to deteriorate when the beam (radiation pattern) is turned. The gain may vary from say 20 at the ideal angle (e.g. 90°) to 12 at another angle (e.g. 60°). Beamsteering tends to increase the number of sidelobes as well. Furthermore, the gain of the antenna may differ by about 10% for the two carrier frequencies; normally the gain is maximized for the receiver carrier frequency. Hence the gain at the transmitting frequency tends to be smaller than the designed value.

7. Study the VSWR of the antenna. For an array antenna, the VSWR of the array should be closer to 1. A single element of the array may have a VSWR of about 2.5.

8. Evaluate the gain-to-temperature (*G/T*) ratio for the antenna. At different elevation angles, the noise will be different. Evaluation of the carrier-to-noise (*C/N*) ratio is also important to determine the characteristics of the electronics required; in general for mobile communication systems, the *C/N* value will vary dramatically from position to position.

9. It is now customary to determine the amount and type of signal processing necessary to make the antenna meet all the required performance criteria. Instead of talking just about antennas, we speak of *adaptive antennas* or *smart antennas*. Hence the signal processing aspects of the antenna should be considered. Signal processing may be performed using smart antennas with signal processing hardware or software.

2   Elementary antenna theory

P.R.P. Hoole

2.1   Introduction

Consider the two Maxwell's equations that couple the electric field intensity **E** and magnetic flux density **B**. The magnetic field intensity $\mathbf{H} = \mathbf{B}/\mu$. The magnetic flux density itself will be expressed as the curl of the magnetic vector potential **A**. A vector is completely defined only after both its curl and divergence are defined. To complete the definition of vector **A**, its divergence will be set to $\nabla\cdot\mathbf{A} = -\mu\varepsilon\,\partial\phi/\partial t$, where ϕ is the scalar potential.

2.1.1   Maxwell's equations

The study of electromagnetic (EM) fields requires the use of the four Maxwell's equations given below:

$$\nabla\times\mathbf{H} = \mathbf{J} + j\omega\varepsilon\mathbf{E}, \tag{1}$$

$$\nabla\times\mathbf{E} = -j\omega\mu\mathbf{H}, \tag{2}$$

$$\nabla\cdot\mathbf{B} = 0, \tag{3}$$

$$\nabla\cdot\mathbf{E} = \rho/\varepsilon. \tag{4}$$

Maxwell's equations relate the two basic physical parameters E (electric field intensity) and B (magnetic flux density) to the sources J (current density) and ρ (electric charge density). In telecommunication, radar and biomedical imaging, the presence and effects of ρ are considered negligible. Thus J is the primary source of information about the voice and video signals being transmitted or the object being imaged.

2.1.2   Vector potential functions

2.1.2.1   The magnetic vector potential **A** for an electric current source **J**
The vector potential **A** is useful in solving for the EM fields generated by an electric current source J. Since the magnetic flux **B** is solenoidal, we have

$$\nabla\cdot\mathbf{B} = 0. \tag{5}$$

Hence **B** can be represented as the curl of another vector because it satisfies the vector identity

$$\nabla\cdot\nabla\times\mathbf{A} = 0, \tag{6}$$

where **A** is an arbitrary vector. Thus we define

$$\mathbf{B}_A = \mu \mathbf{H}_A = \nabla \times \mathbf{A}, \tag{7}$$

where the subscript A denotes that the field is due to the vector potential **A**.
   Substituting eqn (7) into eqn (2) we get

$$\nabla \times \mathbf{E}_A = -j\omega\mu\mathbf{H}_A = -j\omega\nabla \times \mathbf{A} . \tag{8}$$

Rearranging eqn (8) we have

$$\nabla \times \left[\mathbf{E}_A + j\omega\mathbf{A} \right] = 0 . \tag{9}$$

Using the vector identity

$$\nabla \times \left(-\nabla \phi_e \right) = 0 , \tag{10}$$

and comparing eqns (9) and (10), we get

$$\mathbf{E}_A + j\omega\mathbf{A} = -\nabla \phi_e , \tag{11}$$

where ϕ_e represents an arbitrary scalars potential function which is a function of position.
   Taking the curl on both sides of eqn (7) and using the vector identity,

$$\nabla \times \nabla \times \mathbf{A} = \nabla (\nabla \cdot \mathbf{A}) - \nabla^2 \mathbf{A} , \tag{12}$$

we get

$$\nabla \times \left(\mu \mathbf{H}_A \right) = \nabla (\nabla \cdot \mathbf{A}) - \nabla^2 \mathbf{A} . \tag{13}$$

For a homogenous medium eqn (13) reduces to

$$\mu \nabla \times \left(\mathbf{H}_A \right) = \nabla (\nabla \cdot \mathbf{A}) - \nabla^2 \mathbf{A} , \tag{14}$$

which on equating to eqn (1) leads to

$$\mu \mathbf{J} + j\omega\mu\varepsilon\mathbf{E}_A = \nabla (\nabla \cdot \mathbf{A}) - \nabla^2 \mathbf{A} . \tag{15}$$

Substituting eqn (11) into eqn (15) we get

$$\nabla^2 \mathbf{A} + k^2 \mathbf{A} = -\mu \mathbf{J} + \nabla (\nabla \cdot \mathbf{A}) + \nabla \left(j\omega\mu\varepsilon\phi_e \right)$$
$$= -\mu \mathbf{J} + \nabla (\nabla \cdot \mathbf{A} + j\omega\mu\varepsilon\phi_e), \tag{16}$$

where $k^2 = \omega^2 \mu\varepsilon$.
   To define a vector entirely, its curl and divergence should be defined. In eqn (7) the curl has been defined. Now in order to simplify eqn (16), we let

$$\nabla \cdot \mathbf{A} = -j\omega\mu\varepsilon\phi_e . \tag{17}$$

Hence eqn (16) reduces to

$$\nabla^2 \mathbf{A} + k^2 \mathbf{A} = -\mu \mathbf{J} \, , \tag{18}$$

which is known as the Lorentz condition.

Hence we have defined the vector potential $\mathbf{A}$ completely. Also the electric field in eqn (11) can now be written as

$$\mathbf{E}_A = -\nabla \phi_e - j\omega \mathbf{A} = -j\omega \mathbf{A} - j\frac{1}{\omega\mu\varepsilon} \nabla(\nabla \cdot \mathbf{A}) . \tag{19}$$

The electric field $\mathbf{E}$ can now be calculated if the vector potential function $\mathbf{A}$ can be found. The vector potential function $\mathbf{A}$ is now defined in terms of the electric current density $\mathbf{J}$. Consider

$$\nabla^2 A_z + k^2 A_z = -\mu J_z . \tag{20}$$

When the current density is along the z-axis, the vector potential $\mathbf{A}$ is also along the z-axis. At points away from the source, $J_z = 0$. Hence eqn (20) reduces to

$$\nabla^2 A_z + k^2 A_z = 0 . \tag{21}$$

For an infinitesimal source dimension, $\mathbf{A}$ should not be a function of direction (θ, ϕ) and only a function of r, i.e. in a spherical coordinate system $A_z = A_z(r)$ where r is the radial distance. Thus eqn (21) can be written as

$$\nabla^2 A_z(r) + k^2 A_z(r) = \frac{1}{r^2}\frac{\partial}{\partial r}\left[r^2\frac{\partial A_z(r)}{\partial r}\right] + k^2 A_z(r) = 0 . \tag{22}$$

Taking the partial derivative, eqn (22) reduces to

$$\frac{d^2 A_z(r)}{dr^2} + \frac{2}{r}\frac{dA_z(r)}{dr} + k^2 A_z(r) = 0 . \tag{23}$$

This has two independent solutions,

$$A_{z1} = C_1 \frac{e^{-jkr}}{r} , \tag{24}$$

$$A_{z2} = C_1 \frac{e^{+jkr}}{r} , \tag{25}$$

where eqns (24) and (25) represent a wave traveling away and towards the source, respectively. With the source at the origin of the coordinate system we can choose the outwardly traveling wave and hence the solution becomes eqn (25). When $k = 0$ and $\omega = 0$ (static case) the solution becomes

$$A_z = \frac{C_1}{r} . \tag{26}$$

In the presence of the source ($J \neq 0$) and $k = 0$ (static case) the wave equation reduces to

$$\nabla^2 A_z = -\mu J_z.$$ (27)

This form is called Poisson's equation and it has a parallel equation for a scalar potential ϕ, relating it to the charge density ρ:

$$\nabla^2 \phi = -\frac{\rho}{\varepsilon}.$$ (28)

The solution for eqn (28), from Coulomb's law for electrostatics, is given by

$$\phi = \frac{1}{4\pi\varepsilon} \int\!\!\int\!\!\int_v \frac{\rho}{r}\, dv',$$ (29)

where r is the distance from any point on the charge density to the observation point. Since eqn (27) is similar in form to eqn (28), by comparing it to the solution of the second order differential equation, or the Poisson equation, for the scalar potential ϕ, the solution for **A** in general is

$$\mathbf{A} = \frac{\mu}{4\pi} \int\!\!\int\!\!\int_v \mathbf{J}\, \frac{e^{-jkr}}{r}\, dv'.$$ (30)

Hence knowing the current density distribution and the limits of integration, the vector potential function could be found. The procedure to find the electric field E and magnetic field H is given in Section 2.4.

In all our analyses, unless otherwise stated, we shall assume that the current or signal we are dealing with is sinusoidal in the time domain. Hence in the phasor domain analysis we represent the time domain term by $\exp(j\omega t)$. This term gets canceled out on both sides of Maxwell's equations, and hence need not be carried over in antenna analysis. However, when the final complete solution is required, the $\exp(j\omega t)$ term must be attached at the end of the solution. Thus if the final solution looks like $E_0 \exp(-jkR)$, then the complete phasor domain solution is $E_0 \exp(-jkR) \exp(j\omega t)$. Taking the real of this, the time domain solution is $E_0 \cos(\omega t - kR)$. The wave number $k = \omega/c = \omega(\mu_0 \varepsilon_0)^{-1/2}$, where ω, c, μ_0 and ε_0 are the radian signal frequency, velocity of light, free space permeability and free space permittivity, respectively. The values of some important physical constants are given below:

permittivity of free space = $\varepsilon_0 = 8.854 \times 10^{-12}$ F/m;
permeability of free space = $\mu_0 = 4\pi \times 10^{-7}$ H/m;
speed of light in free space = $c = 2.998 \times 10^8$ m/s;
wave impedance in free space $\eta = 120\pi = 376.7\ \Omega$.

Note that even for non-sinusoidal time domain waveforms this analysis is valid, since any time domain signal could be translated into the frequency domain and represented by a summation of sinusoidal waves.

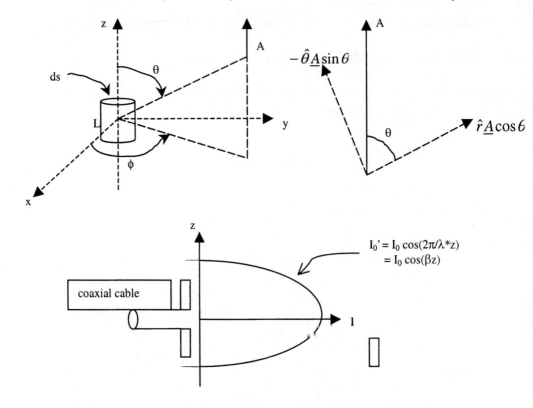

Figure 2.2: Dipole geometry and current distribution.

or sine function shape. Thus this cosine or sinc shaped function needs to be integrated with respect to z for a finite length antenna. Referring to Fig. 2.2, we have

$$A_r = A_z \cos \theta$$

$$= \frac{\mu I L}{4\pi r}e^{-jkr} \cos \theta , \tag{32}$$

$$A_\theta = -A_z \sin \theta = -\frac{\mu I L}{4\pi r}e^{-jkr} \sin \theta , \tag{33}$$

$$A_\phi = 0. \tag{34}$$

When we take the curl of vector potential $\mathbf{A}$ in spherical coordinates, only the $\mathbf{u}_\phi$ component is not zero. Therefore the magnetic field intensity is given by

$$\mathbf{H} = \mathbf{H}_\phi = \frac{1}{\mu}\nabla \times \mathbf{A} = \frac{1}{\mu r}\left(\frac{\partial (rA_\theta)}{\partial r} - \frac{\partial A_r}{\partial \theta} \right)\mathbf{u}_\phi , \tag{35}$$

from which we get

$$\mathbf{H}_\phi = \frac{jkIL}{4\pi}\left[\frac{1}{r} + \frac{1}{jkr^2} \right]\sin \theta \, e^{-jkr}\mathbf{u}_\phi . \tag{36}$$

The exp(−jkr) term is a rotational term in the spatial domain r, similar to the exp(jωt) term in the time domain. Hence the radiated signal rotates in the time and spatial domains. The rotation in the spatial domain, with a period of λ m (where wavelength $\lambda = 2\pi/k$), causes sudden dips in the received signals as observed in mobile communications.

It is seen from eqns (42)–(44) that the infinitesimal current element will generate a magnetic field component that is perpendicular to both the radial and tangential electric field components, respectively.

The E_r field has the radiating near-field and evanescent-field components. These are components that vary inversely as the square and cube of the distance r, respectively. The E_θ field has in addition a component which varies as the inverse of the distance and this is termed the far-field component. In the far-field region E_θ and H_ϕ are connected by a simple coefficient, namely the free space impedance η. This is not true in the near-field region where both E_r and E_θ are related to H_ϕ.

The near-field region is defined as the distance d from the source such that

$$d < \frac{2D^2}{\lambda},\tag{45}$$

where D is the dimension of the antenna along the direction vector. Beyond this distance the radiating near field and evanescent fields disperse very rapidly.

In radar imaging, we use the electric dipole to model the discrete scattering elements. Hence the complete field equations of eqns (42) and (43) become important, in order to determine the dependence of distance on image synthesis in the near-field region. The Doppler frequency shift tends to significantly modify the received power.

2.2.2   *Electric field radiation pattern of an electric dipole*

2.2.2.1   *The E-plane radiation pattern*
The field pattern of an electric dipole in a plane vertical to the xy plane (E-plane) at a distance of 1 m (length of dipole $\lambda/50$, frequency = 100 MHz) is considered.

In Fig. 2.3(a) the variation of the magnitude of the E_r component on an equi-phase surface is shown. In other words, this is the amplitude variation of E_r on a spherical wave front of an electric dipole at a distance of 1 m. It is apparent that the amplitude is dependent on the angle of reception and it is zero at an angle of 90° away from the z-axis. The same observation is true in Fig. 2.3(b), where the nulls are shifted along the z-axis.

Figure 2.3(c) shows the variation of the field amplitude with receiving angle, when the strength is dependent on both E_r and E_θ simultaneously. It is seen that the radiation pattern of the resultant field is almost spherical, like that of a fictitious isotropic or point radiator.

The variation of amplitude due to the resultant signal, as shown in Fig. 2.3(c), is very small for the entire range of reception angle.

These observations are important, in that the angle of reception plays a role in the received signal amplitude and thus prior knowledge on this perturbation is necessary

The electric field intensity may now be obtained from the magnetic field intensity, using Maxwell's equation relation

$$\mathbf{E} = -\frac{j}{\omega \varepsilon} \nabla \times \mathbf{H} \tag{37}$$

Hence

$$\mathbf{E} = \frac{1}{j\omega\varepsilon}\left[\frac{1}{r\sin\theta}\frac{\partial}{\partial\theta}\left(H_\phi \sin\theta\right)\mathbf{u}_r - \frac{1}{r}\frac{\partial}{\partial r}\left(rH_\phi\right)\mathbf{u}_\theta\right] = \mathbf{E}_r + \mathbf{E}_\theta, \tag{38}$$

$$\mathbf{E}_r = \frac{ILe^{-jkr}}{4\pi j\omega\varepsilon}\left(\frac{1}{r\sin\theta}\right)\frac{\partial}{\partial\theta}\left[\sin^2\theta\left(j\frac{k}{r} + \frac{1}{r^2}\right)\right]\mathbf{u}_r$$

$$= \frac{ILe^{-jkr}\cos\theta}{2\pi\varepsilon}\left[\frac{1}{cr^2} + \frac{1}{j\omega r^3}\right]\mathbf{u}_r \ \mathrm{V/m}, \tag{39}$$

$$\mathbf{E}_\theta = \frac{IL}{4\pi j\omega\varepsilon}\left(-\frac{1}{r}\right)\frac{\partial}{\partial r}\left[r\sin\theta\, e^{-jkr}\left(\frac{jk}{r} + \frac{1}{r^2}\right)\right]\mathbf{u}_\theta \tag{40}$$

$$- IL\frac{e^{-jkr}\sin\theta}{4\pi j\omega\varepsilon}\left[-\frac{k^2}{r} + j\frac{k}{r^2} + \frac{1}{r^3}\right]\mathbf{u}_\theta$$

$$= \frac{ILe^{-jkr}\sin\theta}{4\pi}\sqrt{\frac{\mu}{\varepsilon}}\left[j\frac{k}{r} + \frac{1}{r^2} + \frac{1}{jkr^3}\right]\mathbf{u}_\theta \ \mathrm{V/m}, \tag{41}$$

where $\mathbf{u}_r$, $\mathbf{u}_\theta$ and $\mathbf{u}_\phi$ are the spherical coordinate unit vectors. Considering only the terms that are $1/r$ dependent far-field terms, we get

$$\frac{E_\theta(1/r)}{H_\phi(1/r)} = \sqrt{\frac{\mu}{\varepsilon}} = \eta,$$

the intrinsic impedance of the medium containing the dipole, $\eta = 120\pi\,\Omega$ for free space. It should be remembered that to complete the above set of equations the term $\exp(j\omega t)$ should be added to each equation. Hence the complete phase term will be $\exp(j(\omega t - kr))$. The total radiation power cannot vary with distance. Hence power density must vary as $1/r^2$; this rapid decrease in signal strength as the signal propagates in space is due to the spreading out of the antenna beam and not due to power dissipation. In communication design, this power loss is treated as path loss.

$$\mathbf{E}_r = \eta\frac{IL\cos\theta}{2\pi}\left[\frac{1}{r^2} + \frac{1}{jkr^3}\right]e^{-jkr}\,\mathbf{u}_r, \tag{42}$$

$$\mathbf{E}_\theta = j\eta\frac{kIL\sin\theta}{4\pi}\left[\frac{1}{r} + \frac{1}{jkr^2} - \frac{1}{k^2r^3}\right]e^{-jkr}\,\mathbf{u}_\theta, \tag{43}$$

$$E_\phi = 0. \tag{44}$$

to perform corrections on the received signal amplitude. If the modulating factor against angle is known *a priori*, the correction could be incorporated into the signal processing algorithm to automatically correct the received signal strength.

The correction will depend on whether the signal received is due to E_r, E_θ or the resultant and can be implemented using the normalized factors plotted in Fig. 2.3(a), (b) and (c), respectively. For the case where the circularly polarized signal is received, the correction is not necessary since the response is omni-directional.

2.2.2.2   The H-plane radiation pattern
The field pattern of an electric dipole at a distance of 1 m in the horizontal plane (H-plane) (length of dipole $\lambda/50$, frequency = 100 MHz) is considered. The field on the horizontal plane is entirely due to the E_θ component (at $\theta = 90°$, E_r will not be present)

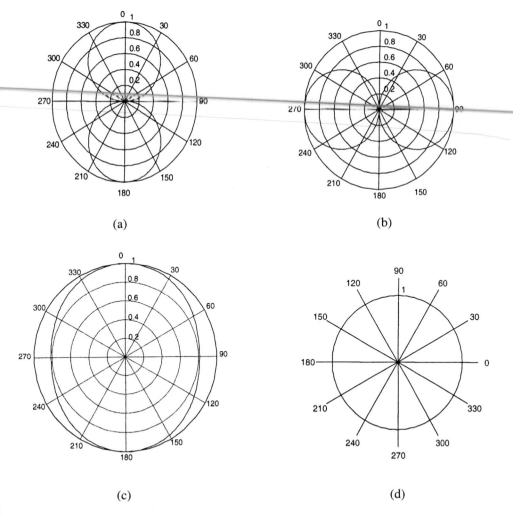

(a)   (b)

(c)   (d)

Figure 2.3: Normalized field pattern (a) of E_r in the vertical plane; (b) of E_θ in the vertical plane; (c) of resultant electric field; and (d) on the horizontal plane.

and it is as shown in Fig. 2.3(d). Amplitude correction is not required as the response is omni-directional.

Far-field or radiation-field components of **E** and **H** are given by

$$H_\phi = j\frac{kILe^{-jkr}\sin\theta}{4\pi r},\qquad(46)$$

$$E_\theta = j\eta\,\frac{kILe^{-jkr}\sin\theta}{4\pi r}.\qquad(47)$$

When we divide eqn (47) by eqn (46), we get the wave impedance,

$$Z = \frac{E_\theta}{H_\phi} = \eta.\qquad(48)$$

Note that the radiation (or far-field) components are the parts of the fields that are dependent on $1/r$, and will dominate the fields at far distances. Thus the E_r components are not significant at far distances. The far-field magnetic and electric field amplitudes decrease as $1/r$ and the phase changes as $-kr$ (i.e. $-2\pi r/\lambda$). The power transmitted by the electric and magnetic fields radiated by an antenna are given by the Poynting vector $\mathbf{P} = \mathbf{E}\times\mathbf{H}$. Therefore the direction of power flow is perpendicular to the plane containing $\mathbf{E}$ and $\mathbf{H}$ It is of interest to note that the $\mathbf{E}_r\times\mathbf{H}$ will result in a power term with a j component; this means that the $\mathbf{E}_r\times\mathbf{H}$ power is an imaginary (or reactive) power term. It is the $\mathbf{E}_\theta\times\mathbf{H}$ power that transfers real (communication or remote sensing) signal power.

$$\text{Power density radiated } P_i = (1/2)\left|\mathbf{E}_{rad}\times\mathbf{H}^*_{rad}\right|$$

$$= (1/2)\left|E_\theta\!\left(\frac{1}{r}\right)H_\phi\!\left(\frac{1}{r}\right)\right| \text{ W/m}^2\qquad(49)$$

ignoring the near-field $(1/r^3)$ and intermediate-field $(1/r^2)$ components.

Expressing this in a more concise form, the radiated power density is given by

$$P_i = \tfrac{1}{2}(\mathbf{E}\times\mathbf{H}^*) = \mathrm{Re}\,\tfrac{1}{2}|H_\phi E_\theta|$$

$$= \frac{1}{2\eta}\left|E_\theta^2\right| \text{ W/m}^2.\qquad(50)$$

Note that radiated power intensity P_i will become half the maximum power intensity radiated at the points on the radiation pattern where electric field intensity is $0.707E_0$, where E_0 is the maximum electric field intensity radiated.

Radiation intensity is given by

$$S = \text{power/unit solid angle}$$

$$= \frac{4\pi r^2 P_i}{4\pi} = r^2 P_i = \frac{\eta}{2}\left(\frac{kIL}{4\pi}\right)^2\sin^2\theta$$

$$\approx r^2 E_\theta^2.\qquad(51)$$

Although we shall elaborate on the near-field phenomena in this chapter, it should be noted that the near-field power flow associated with $\mathbf{E}_r \times \mathbf{H}_\phi$, which is an imaginary (i.e. made of a j-) term, is a reactive power and it circulates around the surface of a fictitious sphere surrounding the antenna. This term is quite significant in the near-field regions of biomedical imaging, synthetic-aperture remote sensing and wireless communications.

Since $H_\phi = E_\theta / \eta$ in the far-field, $P_i = E_\theta^2 / 2\eta$ W/m^2 in the radiation or far-field region. The power density P_i, note, is a function of $\sin^2 \theta$. This means that if we define the directivity of any antenna with respect to the dipole antenna instead of the isotropic antenna, then the $\sin^2 \theta$ term will tend to increase the P_i at any given point over the P_I value of an isotropic antenna with the same input power. Hence the dipole directivity (or gain) of an antenna will be smaller than the isotropic directivity (and gain) of that antenna.

The average power radiated, replacing the wave number k by $k = 2\pi/\lambda$, is obtained by integrating P_i over the spherical surface surrounding the antenna, with the surface element given by $dS = r^2 \sin \theta \, d\theta \, d\phi$.

$$P_r = \frac{1}{2} \frac{I^2 L^2 \sqrt{\mu/\varepsilon}}{4r^2 \lambda^2} \int_{\phi=0}^{2\pi} \int_{\theta=0}^{\pi} \sin^3 \theta \, r^2 \, d\theta \, d\phi$$

$$= \frac{\pi I^2 L^2 \sqrt{\mu/\varepsilon}}{3\lambda^2} \text{ W.} \tag{52}$$

In the above integration, note that by integrating from 0 to π radians in the θ domain and from 0 to 2π radians in the ϕ domain, we integrate over a whole sphere placed around the antenna with its center coinciding with the center of the antenna. In working out the integration, the following rearrangement was made: $\sin^3 \theta = \sin \theta \sin^2 \theta = (1/2) \sin \theta(1-\cos 2\theta) = (1/2) \sin \theta - (1/4)(\sin(-\theta) + \sin 3\theta) = (3/4) \sin \theta - (1/4) \sin 3\theta$. The current I that appears in the power term is the average or uniform current that flows along the antenna. It is equal to the maximum time domain current for short dipoles with a uniform current distribution. If the current distribution was, instead, triangular with maximum current at the center of the dipole with the current tapering off to zero at either ends of the antenna, I should be replaced by $I/2$, and the radiated power is reduced.

Defining the radiation resistance using the circuit relation

$$P_r = \tfrac{1}{2} I^2 R_r \text{ W,} \tag{53}$$

where the current I is the maximum time domain current. Therefore, equating the two expressions we have for average power radiated, we get

$$R_r = \frac{2I^2 L^2 \sqrt{(\mu/\varepsilon)}(\pi/3)}{I^2 \lambda^2}$$

$$= \frac{2}{3\lambda^2} \pi L^2 \sqrt{\frac{\mu}{\varepsilon}} = 80((\pi L)/\lambda)^2 \ \Omega. \tag{54}$$

If the current distribution along the short dipole antenna is triangular in shape, as it usually is, then the current I in eqn (52) should be replaced by $(I/2)$, and the radiation resistance will be $20((\pi L)/\lambda)^2$ Ω.

The power received by a receiving antenna

$$P_R = A_{em} P_i,\tag{55}$$

where A_{em} is the effective aperture of the antenna and P_i is radiation power density at the antenna. The voltage induced along an infinitesimal, short dipole antenna by the radiation electric field E is approximately given by

$$V = EL \text{ V},\tag{56}$$

where E is the electric field associated with the power density that appears at the antenna of length L.

For maximum power transfer the following condition must be achieved by using impedance matching techniques. Ignoring the ohmic resistance of the antenna, the termination impedance must satisfy the following condition:

$$Z_T = R_r - jX_A = Z_A^*,\tag{57}$$

which states that the termination impedance of the cable at the antenna, Z_T, must be equal to the conjugate of the antenna impedance Z_A^*. The antenna impedance is made of the radiation resistance R_r and reactance X_A. Thus the power received by the receiving antenna is given by

$$P_R = \frac{V_{rms}^2}{4R_r} = \frac{E_{rms}^2 L^2}{4R_r} = AP_i = A\frac{E_{rms}^2}{\sqrt{\mu/\varepsilon}},\tag{58}$$

where we have used the matched impedance condition for which the terminating load resistance $R_L (=R_T) = R_r$, and the total terminal impedance $Z_T + Z_A = 2R_r$ so that the power at the receiver is $I_{rms}^2 R_L = (V_{rms}/2R_r)^2 R_L$.

Hence the effective area of the antenna

$$A_{em} = \frac{3\lambda^2}{8\pi} \text{ m}^2$$

or, in general,

$$A_{em} = \frac{D\lambda^2}{4\pi} \text{ m}^2.\tag{59}$$

For a lossless (i.e. ohmic resistance is zero) elemental dipole

$$D = \frac{r^2 \frac{1}{2}(E_\theta H_\phi)_{max}}{P_r/4\pi}$$

$$= \frac{(r^2/2)(I^2 L^2/(4\pi)^2)\sqrt{\mu/\varepsilon}(2\pi/\lambda r)^2}{(\pi/3 I^2 L^2 \sqrt{\mu/\varepsilon})/\lambda^2 4\pi} = \frac{3}{2}.\tag{60}$$

Thus we have

$$\frac{D}{A_{em}} = \frac{3/2}{(3/8\pi)\lambda^2} = \frac{4\pi}{\lambda^2} = \frac{G}{A_{em}}. \tag{61}$$

We have assumed that the receiving antenna is placed at the peak point of the transmitting antenna beam and that the axes of the transmitting and receiving antennas are parallel. In general, if the receiving antenna is placed at an angle θ in the polar coordinate, then $D(\theta) = D \sin^2 \theta$. Furthermore, if the receiver antenna axis is tilted at an angle δ from the transmitting antenna axis, then the effective aperture A_e is no longer equal to the maximum effective aperture A_{em}, but reduced to a value given by $A_e = A_{em} \sin^2 \theta$. Ignoring the losses in the transmitting antenna, we have

$$\text{gain } G = D = \frac{4\pi A}{\lambda^2}; \quad \frac{G_{F1}}{G_{F2}} = \frac{f_1^2}{f_2^2} \tag{62}$$

Note that in much of our discussion we have ignored the ohmic resistance R_o of the antenna, assuming it to be much smaller than the radiation resistance R_r. If both resistances are accounted for, then we have gain $G = (R_r/(R_r + R_o))\ D$; thus the gain G is less than the antenna directivity D.

If we now plot the electric field and magnetic field patterns in the z–y and x–y planes the patterns will be as shown in Fig. 2.4. The three-dimensional radiation pattern (beam) given in Fig. 2.4 shows only the left hand side half of the beam in order to highlight the eight-shaped E-plane pattern. The z–y plane is called the E-plane since the electric field intensity **E** lies in that plane. The x–y plane is called the H-plane since the magnetic field intensity **H** lies in that plane. But it should be remembered that the radiation patterns in both planes apply to the magnitudes of both **H** and **E**, although the two fields will be polarized in different directions. The electric field vector **E** will lie in the E-plane and the magnetic field intensity vector **H** will be in the H-plane. For wire antennas the H-plane pattern is a circle. The three-dimensional structure of the radiation pattern will be shaped like a doughnut. Note that there will be no radiation along the z-axis, the axis on which the antenna is placed. Maximum radiation will be in the x–y plane at $z = 0$. As we move up or down along the z-axis, the field pattern parallel to the x–y plane will gradually decrease. The points at which the field strength is 0.707 of the maximum field strength are called the 3 dB points. In practical antenna engineering, it is generally assumed that radiation in the directions in which the field strength is less than 3 dB will be too weak to observe. Hence the useful radiated power (or effective reception by the same antenna) will be between these two 3 dB points; the angle between the two points is called the half-power beam width (HPBW) of the antenna.

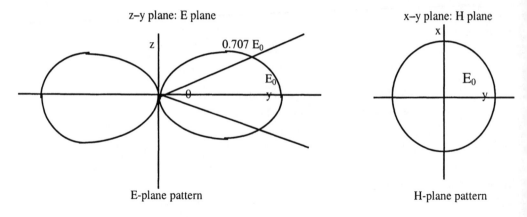

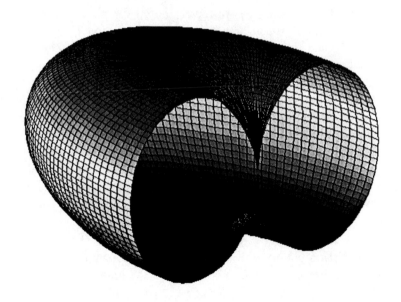

Figure 2.4: Hertzian dipole radiation pattern.

These electrically short antennas are used for special reasons such as when windage resistance should be reduced in high-speed aircraft and warships, when obstacle clearance is a limiting factor, when concealment is required or when an antenna is to be mounted on animals or birds for radio tracking. The radiation resistance of the small antenna is very small, and hence the ohmic resistance has to be accounted for when determining the antenna impedance and radiation efficiency. Impedance matching is critical when using electrically short antennas, since the power captured by such antennas is also quite small and of the order of $(E^2 \lambda^2)/(320 \pi^2)$ W, where E is the incident electric field. When short antennas are to be mounted on vehicles or on the collars of animals, a monopole of length $\lambda/4$ may be placed vertical to an artificial conducting *ground* of length $\lambda/2$ on the collar. The closer an antenna is kept to a human body, the smaller its gain.

2.3   Antenna in motion

When the transmitting or receiving antenna is moving, as in the case of mobile communications (see Fig. 2.5), the received signal frequency is shifted due to the motion of the antenna. Consider a signal $E_0 \cos(\omega t - kR)$ being radiated by an antenna moving directly towards the receiver antenna at a velocity v.

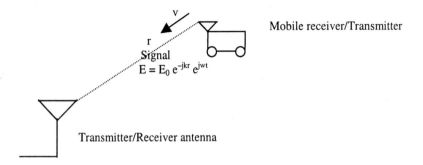

Figure 2.5: Mobile antenna and Doppler frequency shift.

The distance between the antennas is now a function of time, and is given by

$$r = r_0 - vt. \tag{63}$$

Hence the signal picked up by the receiver antenna is

$$E = E_0 e^{j[\omega_0 t - k r_0 + kvt]} \tag{64}$$

Thus we note that the effect of antenna motion is to shift the frequency of the signal. If the transmitter is moving towards the receiver, as in the case we have considered, the signal frequency increases by the Doppler frequency f_D:

$$\omega = \omega_0 + \omega_0 \frac{v}{c} \tag{65}$$

$$f_D = f_0 \frac{v}{c}. \tag{66}$$

Doppler frequency f_D is positive if velocity v is positive and it is negative if the velocity is negative. If the transmitter is moving away from the receiver, the frequency is shifted down. This in turn will modify the power of the received signal. Often some form of compensation is required in communication links to get rid of the effects of the Doppler shift. However in remote sensing systems like radar, the Doppler frequency is treated as a useful parameter from which, for instance, the velocity of the transmitter or target may be estimated. It could be shown that if the transmitter is moving at an angle δ with respect to the straight line connecting the two antennas, the Doppler frequency shift will be

$$f_D = f_0 \frac{v}{c} \cos \delta, \qquad (67)$$

indicating that the maximum frequency shift is when $\delta = 0$. The velocity v is the relative velocity between the transmitter and the receiver; it is positive if the distance r between the transmitter and receiver is getting smaller, and it is negative if r is increasing with time. The Doppler frequency shift tends to significantly modify the received power. From eqn (59) we observe that the effective aperture, and hence the received power, will vary with frequency. From eqn (54) we observe that the radiation resistance, and hence the radiated power, is also frequency dependent. In the fundamental eqns (46) and (47), the wave number $k = \omega/c = 2\pi f/c$, where c is the velocity of light ($=2.998 \times 10^8$ m/s in free space). Thus the magnitudes and phase of the radiated fields are frequency dependent.

2.4   Finite length wire antenna (dipole): the half-wave ($\lambda/2$) dipole

2.4.1   Radiation from an electric dipole antenna of any length L

Consider an antenna of length L, carrying a sinusoidal current I (see Fig. 2.6). We want to find the radiation electric field at point P. We shall consider the observation point to be sufficiently far away (i.e. $r \gg L, \lambda$) to ignore the near- and intermediate-field terms. We shall first work out the electric field intensity for a finite length wire antenna of any length L, and then focus on radiation from the more popular half-wavelength antenna ($L = \lambda/2$). One reason for the popularity of the half-wavelength antenna is that its input impedance is about 73 Ω, which matches well with coaxial cables of line impedance 75 or 50 Ω.

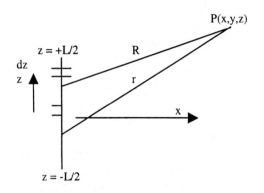

Figure 2.6: General L-length dipole wire antenna.

$$I(z) = \begin{cases} I_m \sin k(L/2 - z), & 0 \le z \le L/2, \\ I_m \sin k(L/2 + z), & -L/2 \le z \le 0. \end{cases} \tag{68}$$

For a short (Hertzian) dipole of length dz carrying current I, the radiation field using eqn (47) is given by

$$dE_\theta = \eta \, dH_\phi = j\eta \frac{kI \, dz \, e^{-jkR}}{4\pi R} \sin \theta \, dz', \tag{69}$$

where

$$R = \sqrt{x^2 + y^2 + (z - z')^2}$$
$$= \sqrt{r^2 + (z'^2 - 2rz' \cos \theta)} \, . \tag{70}$$

The following relations have been used:

$$z \approx r \cos \theta, \qquad R \approx r - z' \cos \theta, \qquad z' \ll r.$$

Therefore the incremental electric field at the observation point due to the small element dz on the dipole antenna may be rewritten as

$$dE_\theta = j\eta \frac{kI e^{-jkr}}{4\pi r} \sin \theta \, e^{jkz' \cos \theta} dz'. \tag{71}$$

Thus the resultant electric field is given by

$$E_\theta = \eta_0 H_\phi = \int_{-L/2}^{+L/2} dE_\theta, \tag{72}$$

yielding

$$E_\theta = \eta_0 H_\phi = j\eta \frac{I_m e^{-jkr}}{2\pi r} \left[\frac{\cos(k(L/2) \cos \theta) - \cos(kL/2)}{\sin \theta} \right]. \tag{73}$$

A sketch of E for different length of dipole will show that the radiation pattern gradually changes as L is increased. When $L = 2\lambda$ the radiation pattern will no longer be maximum in the directions of $\theta = \pi/2, -\pi/2$, but split into four lobes with maximum directions close to $\theta = \pi/4, 3\pi/4, -\pi/4$ and $-3\pi/4$.

2.4.2   Radiation from a half-wave electric dipole antenna: $L = \lambda/2$

A half-wave dipole antenna may be constructed by using two hollow aluminum conductors, each of $\lambda/4$ length. Both elements are vertically aligned with each other, with the transmitter source or the receiver line connected to the two ends adjacent to each other at the center. This is the center fed dipole antenna. Alternatively a $\lambda/4$ length element may be placed over a ground plane. The image acts as the second half of the dipole (see Fig. 2.6).

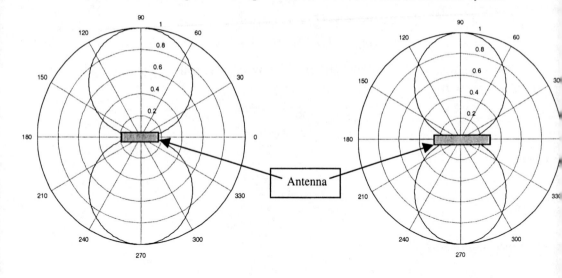

Elemental dipole
$D = G = 1.5$
$f(\theta) = \sin\theta$
(a)

$\lambda/2$-dipole
$D = G = 1.64$
$f(\theta) = \left|\dfrac{\cos((\pi/2)\cos\theta)}{\sin\theta}\right|$
(b)

Figure 2.7: Radiation patterns of dipoles.

Consider now the half-wave dipole antenna shown in Fig. 2.6, with $L = \lambda/2$. The following approximations are made:

$$r_1 = r - z\cos\theta, \tag{74}$$

$\cos\theta \approx$ constant.

Let the current along the antenna be distributed as follows:

$$I = I_0\cos kz. \tag{75}$$

Thus the vector potential at a distance r is given by

$$A_z = \frac{\mu I}{4\pi r}\int_{-\lambda/4}^{+\lambda/4} e^{jkz\cos\theta}e^{-jkr}\cos kz\, dz$$

$$= \frac{\mu I}{4\pi r}\left[e^{j(\pi/2)\cos\theta}\left(0 + k\sin\frac{\pi}{2}\right) - e^{-j(\pi/2)\cos\theta}\left(0 - k\sin\frac{\pi}{2}\right)\right]\frac{e^{-jkr}}{k\sin^2\theta}$$

$$= \frac{\mu I\, e^{-jkr}\cos((\pi/2)\cos\theta)}{2\pi r\, k\sin^2\theta}. \tag{76}$$

In spherical coordinates, we have

$$A_r = \mu I \ e^{-jkr} \ \frac{\cos((\pi/2)\cos\theta)\cos\theta}{2\pi r \ k \ \sin^2\theta} \ , \tag{77}$$

$$A_\theta = \frac{-\mu I \ e^{-jkr} \ \cos((\pi/2)\cos\theta)}{2\pi r \ k \ \sin\theta} \ , \tag{78}$$

$$A_\phi = 0 \ . \tag{79}$$

The radiation (i.e. $1/r$) components of the magnetic field and electric field may be obtained from the vector potential. The radiation fields are given below:

$$H_\phi = jI \ e^{-jkr} \ \frac{\cos((\pi/2)\cos\theta)}{2\pi r \ \sin\theta} \ \text{A/m} \ , \tag{80}$$

$$E_\theta = jkI \ \frac{e^{-jkr} \ \cos((\pi/2)\cos\theta)}{2\pi\varepsilon\omega r \ \sin\theta} \ \text{V/m} \ . \tag{81}$$

Average power density radiated

$$P_i(\theta) = \frac{I^2 k \ \cos^2((\pi/2)\cos\theta)}{8\pi^2\varepsilon\omega r^2 \ \sin^2\theta} \left(= \tfrac{1}{2}|E_{\theta\text{rad}}||H_{\phi\text{rad}}| \right)$$

$$= \frac{I^2\sqrt{\mu/\varepsilon} \ \cos^2((\pi/2)\cos\theta)}{8\pi^2 r^2 \ \sin^2\theta} \ \text{W/m}^2 \ . \tag{82}$$

Hence the average power radiated

$$P_{av} = \int_0^{2\pi}\int_0^{\pi} P_i(\theta)r^2 \ \sin\theta \, d\theta \, d\phi$$

$$= \frac{2\pi I^2\sqrt{\mu/\varepsilon}}{4\pi} \int_0^{\pi} \frac{\cos^2((\pi/2)\cos\theta)}{\sin\theta} \, d\theta \ \text{W}$$

$$= \frac{1.218\sqrt{\mu/\varepsilon} \ I^2}{4\pi} = \frac{1}{2}I^2 R_r \ \text{W} \ . \tag{83}$$

Hence the radiation resistance of a half-wave dipole is given by

$$R_r = \frac{2(1.218)(120\pi)}{4\pi} = 73\,\Omega \ . \tag{84}$$

Therefore the standard 75 Ω coaxial cable may be very closely matched to a half-wave dipole antenna without any additional matching circuits, which accounts for the popularity of the half-wave dipole antenna. The reactive impedance $X = \text{j}42.5\,\Omega$ if $L < \lambda/2$.

The maximum power radiated/unit solid angle is $r^2 P_{av}(\theta)$ with $\theta = 90^{\circ}$. The gain

$$G_{\lambda/2} = \frac{4\pi r^2 (1/2)\left(kI^2 / 4\pi^2 (\varepsilon\omega) r^2\right)/120\pi}{\left[1.218(120\pi)I^2 / 4\pi\right]}$$

$$=1.64. \tag{85}$$

Thus the maximum gain (or directivity) of a half-wave dipole is 1.64, which is greater than the figure of 1.5 we had for the infinitesimal (or Hertzian) dipole.

In general, for a lossless half-wave dipole antenna, the direction (θ) dependent directivity and gain are given by

$$(\theta) = G(\theta) = 1.64\left(\left|\frac{\cos\left((\pi/2)\cos\theta\right)}{\sin\theta}\right|\right)^2. \tag{86}$$

The effective aperture of the half-wave dipole antenna, with $D = 1.64$, is given by

$$A_{em} = \frac{1.64}{4\pi}\lambda^2 = 0.13\lambda^2. \tag{87}$$

The radiation patterns for an elemental dipole (Fig. 2.7(a)) and a half-wave dipole (Fig. 2.7(b)) are shown. Notice that the half-wave dipole has a narrower beam. Since $z = 0$ is a plane of symmetry for the half-wave dipole current (as well as some other wire antennas) the half-wavelength long ($\lambda/2$) wire may be replaced by a quarter-wavelength ($\lambda/4$) wire placed above a ground plane. The radiation pattern of the $\lambda/4$-length antenna will be the same as that of a $\lambda/2$-length wire antenna. The directivity will be doubled, but the radiated power, radiation resistance and the input impedance will all be halved for a $\lambda/4$-length antenna placed above a ground plane. The ground plane can be made of six or more horizontally placed, grounded wires of length $\lambda/2$ arranged in a spider-web like pattern just under the quarter-wave antenna wire.

Example 1. The antenna wire radius = 0.4 mm. A 0.5 m long dipole is operating at 300 MHz. The conductivity of the antenna is given by $\sigma = 6 \times 10^7 \ \Omega^{-1} m^{-1}$. Discuss its radiation efficiency.

$$\text{Skin depth } d = \sqrt{2/\omega\mu\sigma}. \tag{88}$$

The ohmic resistance, for radius r and length L,

$$R_o = L/\sigma A = L/\sigma(2\pi rd). \tag{89}$$

Radiation resistance, since $L = 0.5$ m $= \lambda/2$ ($\lambda = 3 \times 10^8/300 \times 10^6 = 1$),

$$R_r = 73.2 \ \Omega.$$

Radiation efficiency

$$\eta_{R1} = G/D \approx R_r/(R_o + R_r). \tag{90}$$

If frequency $= 3.0$ MHz, $\lambda = 100$ m $>> L$, then the antenna is an infinitesimal (Hertzian) dipole

$$R_r = 80(\pi L/\lambda)^2,$$

and the new efficiency

$$\eta_{R2} << \eta_{R1}.$$

At microwave frequencies dielectric rods (called dielguides) of relative permittivity of the order of 4 (and conductivity of the order of $10^{-7}\,\Omega^{-1}\mathrm{m}^{-1}$!) are used to direct the microwave signals to the parabolic reflector. The ohmic loss becomes too large for metallic conductors to be used at microwave and millimeter frequencies. However, as in the case of optical signals moving along a glass optical fiber, microwave and millimeter waves tend to cling on to dielectric rods, just as low frequency signals tend to cling onto metallic conductors. Thus the metallic dipole antennas are mostly used for frequencies of the order of 1 GHz and lower; at higher frequencies we need the hollow metallic or dielectric waveguides to construct effective antennas. Sometimes metallic dipole antennas are coated with low conductivity, high permittivity dielectric coats to increase antenna capacitance and hence antenna bandwidth, although this results in a reduction of radiation efficiency.

Example 2. Two antennas of 0.5 m length are operating at 300 MHz. One antenna radiates 500 W. The second antenna is 1 km away, at $\theta = 90°$, $\phi = 60°$. The axes of both antennas are parallel to each other. Determine the power received by the second antenna and the current induced in it.

The incident power density is given by (see Fig. 2.8)

$$P_i = \frac{1}{2\eta}|E_\theta|^2, \quad \text{where } \eta = 120\pi.$$

The radiated power from the transmitter $P_r = I_{rms}^2 R_r = \frac{1}{2}I^2 R_r$, where I_{rms} is rms current and I is maximum current.

$$I = \sqrt{\frac{2P_r}{R_r}} = \sqrt{\frac{2\times 500}{73.2}} = 3.7 \text{ A}.$$

The radiation electric field

$$|E_\theta| = \frac{\eta I}{2\pi r}\left|\frac{\cos(k(L/2)\cos\theta) - \cos k(L/2)}{\sin\theta}\right|,$$

$\theta = \pi/2$, $kL/2 = \pi L/\lambda = \pi/2$; the length of the antenna $L = 0.5$ m, $\lambda = 1$ m. We have cos $(kL/2) = 0$. The receiver is at $r = 1000$ m.

$$|E_\theta| = \frac{120\pi I}{2\pi r}\left|\cos\left(\frac{\pi}{2}0\right)\right| = (60/r)I = 0.221 \text{ V/m.}$$

The incident power on the receiver antenna is $P_i = (1/2)E^2/120\pi$ W/m^2 = 0.065 mW/m^2.

Hence the total power received

$$P_R = P_i A_{em} = \frac{1}{2\times120\pi}(0.221)^2 \times 0.13 \times 1 \text{ W} = 8.4 \times 10^{-6} \text{ W.}$$

Recalculate P_R for (i) $\theta = \pi/4$, (ii) $\theta = \pi/6$ and (iii) $f = 1280$ MHz, $L = \lambda/2$ and $\theta = \pi/2$. The problem may be solved using the Friis equation as well. The receiver power is generally very small, and of the order of μW, nW or pW. Hence efficient low noise amplifiers (LNA) are required to keep SNR in receiver electronics to acceptable values.

In the MF band (960, 1200 and 1400 kHz) 10 kW transmitters radiating signals from a dipole antenna at a height of 0.25λ will provide a radio coverage within a 140 km radius of ground wave service. Ground waves, as opposed to sky waves, are electromagnetic waves radiated by the antenna that skims along the surface of the earth, the most common mode of transmission at LF and MF bands. Due to losses in the ground, the radiation efficiency of the antenna could go down from 95% for a ground conductivity of 0.01 S/m to about 80% when the ground conductivity is 0.001 S/m. For higher frequency transmission the ground wave becomes an ineffective transmission route and the sky waves should be used as in the case of cellular communications. Furthermore, in order to prevent distortion of the antenna radiation pattern by obstructions, the antenna should be installed with a clearance of 10 m if the area of the obstruction is 300 m^2, and a clearance of 80 m should be allowed if the obstruction area is about 2500 m^2. HF dipole antennas transmitting at 100 kW may provide coverage of an entire country if frequencies of 7.3, 5.4 and 3.4 MHz are used to combat the variations in the ionosphere height (approximately 300 km above the earth due to diurnal, seasonal and the 11-year sunspot cycles). In this case the antennas will have to be very long, and the radiated electromagnetic signal bounces off the ionosphere and is reflected to various regions of the country.

2.5   Radiation resistance

A few points regarding the radiation resistance may help at this stage. The radiation resistance is not the ohmic resistance of the antenna conductor; it is associated with the power radiated out from the antenna. The ohmic resistance is related to the power dissipated or lost in the antenna conductor. The ohmic resistance is the (power loss in the antenna)/I^2 whereas radiation resistance is equal to (power radiated)/I^2, where I is the current along the antenna. The resistances of half-wave ($\lambda/2$) and quarter-wave ($\lambda/4$) antennas are 73 and 36 Ω; the radiation resistance is a function of the length of the antenna. As the dipole antenna length is increased the radiation resistance tends to oscillate in between 70 and 130 Ω for different lengths of the dipole; however at antenna lengths of 80λ and above the resistance tends to settle down at about 130 Ω. Antennas like the Marconi antenna are half-wave antennas, where only a quarter-wavelength long conductor is placed above the ground; the reflection (image) in the ground then makes it a half-wave antenna, but the input resistance of the Marconi antenna is only 36 Ω.

The height at which the antenna is mounted also has an effect on the radiation resistance. If a half-wave ($\lambda/2$) antenna is placed vertical above the ground, the radiation resistance will be close to 73 Ω at all heights above 0.5λ. Below this height, the radiation resistance tends to increase exponentially. This could cause problems with impedance matching with the coaxial cable connected to the antenna. If the half-wave antenna is placed horizontal above the ground, the radiation resistance tends to decrease rapidly towards zero at heights below 0.5λ. At heights above this, the radiation resistance of a horizontally placed half-wave dipole also tends to be close to 73 Ω. It is also useful to note that the electrical length of the antenna is generally about 5% greater than the physical length of the antenna, especially when the antenna conductor is not very thin. Hence to construct a half-wave antenna at 200 MHz ($\lambda = 1.5$ m), the physical length of the antenna should be $0.95 \times 0.5\lambda = 0.7125$ m. It is very difficult to get an exact match between the physical and electrical lengths. If the physical length of the antenna is slightly longer than the electrical length, the antenna will be inductive (use a series capacitor at the antenna input to cancel out the inductive effect); if the physical length is shorter than the electrical length, then it will be capacitive (use a series inductor to cancel out the capacitive effect).

2.6   Impedance matching

To improve the transmission line–antenna frequency characteristics, it is important to ensure that the impedances of the line and antenna are matched. Else, where there is impedance mismatch, part of the power delivered to a transmitting antenna will be reflected back to the source. In the case of a receiver antenna, all the power captured by the antenna P_T will not be delivered to the receiver electronics if the impedance of the antenna and that of the line/low noise amplifier (LNA) combination are not properly matched. When a signal is going from a cable of impedance Z_1 to an antenna of impedance Z_2, the amount of reflected power is $((Z_2 - Z_1)/(Z_1 + Z_2))^2$. A variety of matching techniques are available, the quarter-wave transformer being one of the most popular methods to match an open wire line to a half-wave dipole antenna. As shown in Fig. 2.8, a short circuited line (short circuit stub) is connected at the point where the line (or cable) is connected to the antenna such that the impedance at the antenna terminal is transformed from Z_0 to $Z_1 = (Z_0 R_{in})^{1/2}$, where R_{in} is the total impedance of the antenna (see Fig. 2.9).

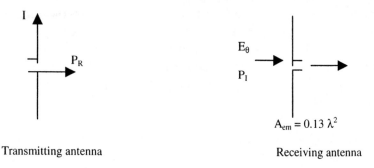

Transmitting antenna   Receiving antenna

Figure 2.8: Transmitting and receiving half-wave dipole antennas.

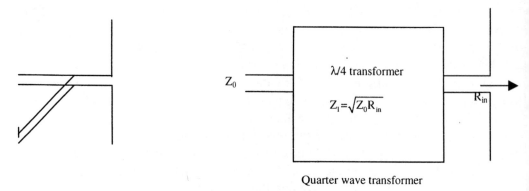

Quarter wave transformer

Figure 2.9: Impedance matching.

The signal-to-noise ratio (SNR) will be greatly increased where impedances are properly matched. Consider the antenna as receiving a signal $E+E_n$, where E is the desired signal at frequency f_0 and E_n is noise. The antenna and the line/LNA impedances are matched at frequency f_0 such that signal E is captured efficiently. If E_n is additive white Gaussian (random) noise (AWGN), it will have a frequency spread over the entire bandwidth of the antenna. Now by matching the circuit at f_0, we inevitably reduce the strength of E_n delivered to the LNA, since only the portion of E_n at f_0 will be entirely transferred to the LNA. The other power components of E_n will be reflected back to the antenna. Hence the SNR at the LNA is improved. However, note, if the noise is Rayleigh, then E_n is also at f_0; in the case of Rayleigh noise, common in wireless communications, E_n is a reflected part of E. Hence for Rayleigh noise, an improvement of SNR should not be expected with impedance matching, although impedance matching is required to capture the desired signal E.

2.7   Radiation safety

One important aspect to address when installing an antenna is the radiation level. Consider a mobile phone that transmits a signal at $P_r = 1$ W. Then the power density at a distance of 2 cm from the antenna, optimistically assuming isotropic radiation, is given by $P_r/4\pi r^2 = 1/4\ \pi(0.02)^2 = 20$ mW/cm^2, a figure which should be multiplied by the directivity (e.g. $D_{\lambda/2} = 1.64$) of the antenna in order to get a more accurate value. Is this power density within the safety limits for a human head that is close to the mobile phone antenna? It is the responsibility of the telecommunications authority to ensure that all transmitters confine to the limits set on the maximum radiation electric field intensity permitted or the maximum radiation power intensity permitted. Biological tissues have electrical properties that are frequency dependent. From Table 2.1, note that the conductivity of most materials increases with signal frequency. This means that as higher communication or remote sensing signals impinge on a human body, relatively more energy will be absorbed. More absorption indicates that more power is converted into heat loss, which could lead to the burning of tissues. Dead tissues are associated with cancer. Table 2.2 shows the impedances of various biomaterials. The complex impedance values of the biomaterials are calculated using the conductivity and relative permittivity values given in Table 2.1. These impedances determine the amount of radiation signals which penetrate into particular materials, since the transmission coefficient of material interface is given by $2Z_2/(Z_1+Z_2)$, where the signal is assumed to be going from a medium of intrinsic impedance Z_1 into a medium of impedance Z_2.

Table 2.1: Dielectric properties of biomaterials at different frequencies.

Material	27.12 MHz		63 MHz		350 MHz	
	σ	ε_r	σ	ε_r	σ	ε_r
Muscle	0.75	106	0.93	88	1.33	53.0
Blood	0.28	102	–	–	1.20	65.0
Skin	0.74	106	–	–	0.44	17.6
Brain	0.45	155	0.55	109	0.65	60.0
Fat	0.04	29	0.06	11.6	0.07	5.7

Table 2.2: Electrical impedance of biomaterials.

Frequency (MHz)	Muscle	Blood	Skin	Brain	Fat
27.12	16.7	26.0	16.8	20.6	60.2
	39°	31°	39°	31°	21°
63.00	22.5	–	–	27.3	84.6
	36°			28°	27°
350.00	40.5	39.8	70.4	45.5	144.6
	26°	22°	26°	15°	17°

In Table 2.3 the radiation limits applied in general are shown. From country to country these values may differ. Some authorities, like the former Soviet Union telecommunica-   tions authority, impose stricter limits. The maximum permissible electric field at extremely low frequencies used to be 25 V/m (rms) in Russia instead of the 87 V/m shown in Table 2.3; at microwave communication frequencies, the limit set was 0.01 mW/cm^2 instead of 10 mW/cm^2.

Table 2.3: Radiation limits (general public).

Frequency (MHz)	E (rms)	H (rms)	Power (W/m^2)
0.01	87	$0.23/\sqrt{f}$	
1–10	$87/\sqrt{f}$	$0.23/\sqrt{f}$	
10–400	27.5	0.073	2
400–2000	$1.375\sqrt{f}$	$0.0037\sqrt{f}$	$f/200$
2000–300,000	61	0.16	10 (mW/cm^2)

International Radiation Protection Association.

Depending on the regulations imposed by the telecommunications authorities, it is the duty of the antenna systems design engineer to ensure that the maximum radiation field from the antenna does not exceed the specified maximum limit. Therefore electric field strength measurement very close to the antenna is necessary to ensure that these limits are observed. This is one important reason why in wireless communications many base station antennas are required to provide service: using one base station antenna to serve an area larger than, say, 5 km^2 area will mean that the electric field radiated will need to be stronger to transmit clear signal to far distances. This may prove unsafe for people close to the base station (BS) of a wireless communication system.

2.8   The effect of antenna height and ground reflection

Consider the case where wire antennas are being used in wireless communications. The stationary, base station (BS) antenna will be placed on top of a building or tower of height, say, h_1 relative to the ground. The mobile phone or mobile station (MS) antenna will be closer to the ground at a height, say, h_2 relative to the ground. When a signal is being radiated from the MS antenna to the BS antenna, ignoring reflections due to all other objects except for the ground, two signals will arrive at the BS antenna for each transmission. Signal 1 will be the direct signal traveling from the MS antenna to the BS antenna, and signal 2 will be the ground-reflected signal which travels from the MS to the ground and is then reflected back to the BS antenna.

Let the reflection coefficient of the ground for vertically polarized signals be denoted by Γ_v. Then the resultant signal appearing at the BS antenna is given by

$$E = (k/r)E_0 e^{j\omega t} + \Gamma_v (k/r)E_0 e^{j(\omega t + \Delta\delta)} \tag{91}$$

or

$$E = (k/r)E_0 e^{j\omega t} (1 + \Gamma_v e^{j\Delta\delta}), \tag{92}$$

where $\Delta\delta = k\Delta r$ is the phase difference due to the extra distance Δr that the reflected signal has traveled. The (k/r) factor appears from the forms that we got for radiation fields in Sections 2.3 and 2.4. If the direct distance between the MS and BS antennas is r, then

$$\Delta r = ((h_1+h_2)^2 + r^2)^{1/2} - ((h_1-h_2)^2 + r^2)^{1/2}, \tag{93}$$

which for $r \gg h_1+h_2$ reduces to $\Delta r = (2h_1h_2)/r$.

Hence the resultant electric field signal picked up at the BS antenna is given by

$$E = (k/r)E_0\, e^{j\omega t}\, (1 + \Gamma_v\, (\cos \Delta\phi + j \sin \Delta\phi)) \tag{94}$$

or

$$E \cong (k/r)E_0\, e^{j\omega t}\, (1 + \Gamma_v + j\Gamma_v\Delta\phi). \tag{95}$$

For small values of $\Delta\phi$, $\Delta\phi = k\Delta r = k(2h_1h_2)/r$.

The (conductivity σ, relative permittivity ε_r) values of seawater (80, 4), rural earth (14, 0.002), urban ground (3, 0.0001), turf (5, 0.01) and dry sandy loam (3, 0.03) are such that in wireless communications, the reflectivity coefficient of ground is close to unity at UHF frequencies. For ground with reflection coefficient $\Gamma_v = -1$, $E \cong -j(k/r)E_0\, e^{j\omega t}\, \Delta\phi = -j(2h_1h_2)\, (k^2/r^2)E_0\, e^{j\omega t}$ V/m. Therefore the power density (in W/m^2) received is given by

$$P_i = E^2/2\eta = 2(h_1h_2)^2(k^4/r^4)(E_0^2/\eta). \tag{96}$$

Therefore we note that in wireless, mobile communication antennas, the radiated electric field varies as $1/r^2$ instead of $1/r$ with distance, and the power varies as $1/r^4$ instead of $1/r^2$ due to the adverse effects of the ground-reflected path signal interfering with the direct signal. In order to reduce the fading due to the reflected path, more directive beam antennas could be used in order to avoid aiming signals towards the ground. To get such directive radiation beams, we need to use more than one antenna element to achieve a smart antenna with a beam adaptively being changed as the MS antenna keeps moving in the spatial domain.

2.9   Antenna radiation in the near-, intermediate- and far-field regions

In contemporary imaging algorithms the reflected or re-radiated electromagnetic signals are assumed to be generated in the far-field region of the transmitting/receiving antenna system. This is feasible by modeling the radar scatterers as infinitesimal dipoles instead of the usual point scatterer model. According to eqn (44) the far-field condition is satisfied at even a few wavelengths away from the source.

For the synthetic aperture antenna, found in synthetic aperture radar (SAR) for instance, scenario eqn (44) has to be modified as

$$d < \frac{2D_{\text{eff}}^2}{\lambda}, \tag{97}$$

where D_{eff} is the effective aperture of the source antenna/element. D_{eff} is the dimension of a real aperture, which would give the same beam width as the synthetic aperture. This increases the near-field distance many-fold and thus focusing becomes imperative in imaging routines that synthesize a narrow beam width using SAR principles.

2.9.1 Magnitude modifying factor

In the near-field region as defined by eqn (97) it is important to consider the variation of the magnitude of the return signal with distance. This is because in the near-field region the magnitudes of the electric and magnetic field components are inversely proportional to first, second and third order of the distance in general. In addition, two orthogonally polarized field components are present: E_r, E_θ in the case of the electric dipole and H_r, H_θ in the case of the magnetic dipole. By comparison, in the far-field region of the electric or magnetic dipole only E_θ or H_θ are present, respectively. Thus in the near-field region there exist additional factors in the return signal which had hitherto not been considered in imaging routines. By modeling the return signals as produced by electric dipoles and magnetic dipoles as in radar and MR imaging respectively, these additional factors may be included in the electromagnetic signal model.

The near-field effect manifests itself by modifying the magnitude and phase of the return signal. Hence these factors have been named the magnitude modifying factor (MMF) and the phase modifying factor (PMF), respectively (Naveendra and Hoole, 1999).

As there is equivalence in the field equations of E_r, H_r and E_θ, H_θ, these two factors MMF and PMF may be analyzed in terms of the electric field components only. Details of imaging in the presence of MMF and PMF are presented in Chapter 4. This will equally apply to the magnetic field components apart from the change in coefficients. The extension to magnetic fields in magnetic resonance imaging will be addressed in Chapter 6.

The magnitude modifying factors due to E_r and E_θ are given by eqns (42) and (43), respectively. It is a function of the distance between the source and the observation point. The received signal strength A is defined in Chapter 4. The MMFs associated with the resultant of E_r and E_θ are defined in eqns (48) and (49), respectively.

$$F_R(r) = A \left(\frac{1}{r^4} + \frac{1}{k^2 r^6} \right)^{1/2}, \tag{98}$$

$$F_\theta(r) = A \left(\frac{1}{r^2} - \frac{1}{k^2 r^4} + \frac{1}{k^4 r^6} \right)^{1/2}. \tag{99}$$

Hence the resultant MMF is given by

$$F(r) = \sqrt{F_R^2 + F_\theta^2}. \tag{100}$$

The MMFs related to E_r, E_θ and the resultant field are plotted in Fig. 2.10. The range of distance used here to compute the MMFs corresponds to applications in near-field radar imaging, near-field RCS measurements and biomedical imaging. The variation of the MMF due to E_r is rapid in the region of 0.5–1.0 m. A difference in distance of a few tens of centimeters between two equal scattering cross sections will modify the scattered signal strength between the two surfaces by a factor of 2 or more. Comparatively, the variation of the MMF due to E_θ is not rapid and the degree of change is not marked. The variation of the resultant ($\sqrt{(F_r^2 + F_\theta^2)}$) more resembles the variation due to E_r. Since the thesis explores the possibility of imaging with E_r, E_θ and $\sqrt{(F_r^2 + F_\theta^2)}$, the latter being the term used in current imaging technology, all modifying factors are included in the imaging algorithms proposed.

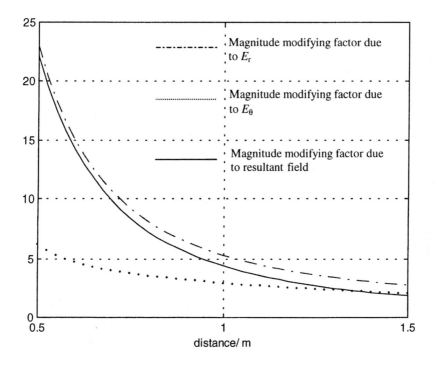

Figure 2.10: Variation of magnitude modifying factor.

2.9.2   Phase modifying factor

Due to the complex nature of the electromagnetic signals in the near-field region, they give rise to a distance dependent phase angle. This has been termed the phase modifying factor (PMF). The phase angles in the signals can be calculated from eqns (42) and (43) when the signals are modeled by E_r and E_θ, respectively.

$$\phi_r = \tan^{-1}\left(-\frac{1}{kr}\right),$$
(101)

$$\phi_\theta = \tan^{-1}\left(\frac{k^2 r^2 - 1}{kr}\right).$$
(102)

The variation of this additional phase angle is shown in Fig. 2.11. We have discounted the distance dependent phase shift $(-kr)$, which is normally filtered out by the standard matched filter.

The variation of phase associated with E_θ is more pronounced than that of the E_r PMF. In digital communication systems using phase-shift-keying (PSK), SAR and MRI, the phase information is important. Knowledge of the underlying signal model will help in evaluating the additional phase angle and correcting for it.

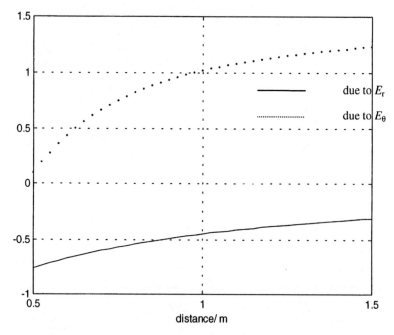

Figure 2.11: Variation of the phase modifying factor.

2.9.3   Inverse Doppler effect in the near-field region

When there is relative motion between the transmitter and the target scatterer, the frequency observed at the receiver is different from that which is transmitted. This is called the Doppler phenomenon. The Doppler frequency shift in the near-field region is different from that experienced by an observer in the far-field region since the Doppler frequency shift depends on the distance between the source and the observation point. Hence the radial and tangential electric field (magnetic field) components of the electromagnetic radiation exhibit different Doppler frequency shifts. The Doppler frequency shift of the radial electric field component is given by

$$\omega_{dr} = \frac{k^2 v(r_0 - vt)}{[1 + (kr_0 - kvt)^2]}. \tag{103}$$

The Doppler frequency shift of the tangential electric field component is given by

$$\omega_{dt} = \frac{k^3 v[k^2(r_0 - vt)^2 - 2](r_0 - vt)^2}{[k^4(r_0 - vt)^4 - k^2(r_0 - vt)^2 + 1]}, \tag{104}$$

where k is the wave number, r_0 is the initial distance of the scatterer from the transmitter and v is the relative velocity between the transmitter and the scatterer. Figure 2.12 shows the variation of the Doppler frequency of the radial electric field component when a scatterer, which is initially at a distance of 100 m, moves past the transmitter at a velocity of 100 m/s.

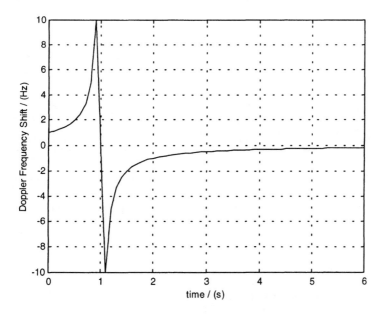

Figure 2.12: Doppler frequency shift.

From Fig. 2.12, for $t < 1$ s, it is seen that when the scatterer is approaching the transmitter the Doppler frequency should increase. But from Fig. 2.12 it is also seen that, very close to the transceiver ($0.9 \leq t \leq 1$), the Doppler frequency shift is actually decreasing. This is the so-called inverse Doppler effect. A similar effect is seen in the time interval $1 \leq t \leq 1.1$. The variation of the maximum Doppler frequency shift with velocity of the radial electric field component is shown in Fig. 2.13. From Fig. 2.13 it is seen that the maximum Doppler frequency shift of the radial electric field component varies erratically with velocity.

A similar inverse Doppler effect is exhibited by the tangential electric field component as well. In imaging of moving objects in the near-field region, this inverse Doppler effect has to be taken into consideration in the imaging routines. In this thesis, the relative velocity between the transceiver and scatterers was assumed to be zero.

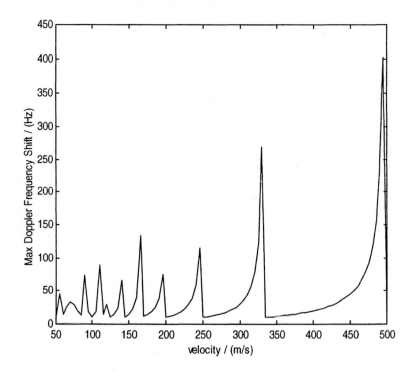

Figure 2.13: Variation of maximum Doppler frequency with velocity.

2.10   The magnetic dipole: the loop antenna

2.10.1   The electric vector potential F for magnetic current source M

Using the volume or surface equivalence theorem, a hypothetical magnetic current is invoked to analyze fields radiated from magnetic current source M. In a homogenous region with $\mathbf{J} = 0$ and $\mathbf{M} \neq 0$, we have $\nabla \cdot \mathbf{D} = 0$. Hence $\mathbf{E}_F$ should satisfy

$$\mathbf{E}_F = -\frac{1}{\varepsilon}\nabla\times\mathbf{F}. \tag{105}$$

Substituting eqn (105) into eqn (1) we get

$$\nabla\times(\mathbf{H}_F + j\omega\mathbf{F}) = 0. \tag{106}$$

Using vector identity eqn (6), eqn (106) yields

$$\mathbf{H}_F = -\nabla\phi_m - j\omega\mathbf{F}, \tag{107}$$

where an arbitrary magnetic scalar potential function ϕ_m is introduced. Taking the curl of eqn (105), we have

$$\nabla\times\mathbf{E}_F = -\frac{1}{\varepsilon}\nabla\times\nabla\times\mathbf{F} = -\frac{1}{\varepsilon}\left[\nabla(\nabla\cdot\mathbf{F}) - \nabla^2\mathbf{F}\right]. \tag{108}$$

Comparing eqn (108) with eqn (1), but using a magnetic current source, we have

$$\nabla\times\mathbf{E}_F = -\mathbf{M} - j\omega\mu\mathbf{H}_F, \tag{109}$$

which leads to

$$\nabla^2\mathbf{F} + j\omega\mu\varepsilon\mathbf{H}_F = \nabla(\nabla\cdot\mathbf{F}) - \varepsilon\mathbf{M}. \tag{110}$$

Substituting eqn (194) into eqn (87) results in

$$\nabla^2\mathbf{F} + k^2\mathbf{F} = -\varepsilon\mathbf{M} + \nabla(\nabla\cdot\mathbf{F}) + \nabla(j\omega\mu\varepsilon\phi_m). \tag{111}$$

By setting

$$\nabla\cdot\mathbf{F} = -j\omega\mu\varepsilon\phi_m, \tag{112}$$

eqn (111) reduces to

$$\nabla^2\mathbf{F} + k^2\mathbf{F} = -\varepsilon\mathbf{M}. \tag{113}$$

Thus the vector potential $\mathbf{F}$ is completely defined in terms of the curl and divergence equations of eqns (105) and (112), respectively.

By analogy with eqn (30) and changing the parameters suitably, the solution for the electric vector potential $\mathbf{F}$ can be defined in terms of the magnetic current $\mathbf{M}$ as

$$\mathbf{F} = \frac{\varepsilon}{4\pi}\iiint_v \mathbf{M}\frac{e^{-jkr}}{r}\,dv'. \tag{114}$$

2.10.2   Field equations for a circular current loop

The potential function due to the current I in the loop as shown in Fig. 2.14 is defined by

$$A(x, y, z) = \frac{\mu}{4\pi} \int_c I_e(x', y', z') \frac{e^{-jkR}}{R} \, dl' , \qquad (115)$$

where R is the distance from any point on the loop to the observation point and dl' is an elemental length of the conductor.

The current $I(x', y', z')$ could be written as

$$I_e(x', y', z') = \hat{a}_x I_x(x', y', z') + \hat{a}_y I_y(x', y', z') + \hat{a}_z I_z(x', y', z'). \qquad (116)$$

Since current flow is confined to I_ϕ, eqn (116) could be written in the spherical coordinate system as

$$I_e = \hat{a}_r I_\phi \sin\theta \sin(\phi - \phi') + \hat{a}_\theta I_\phi \cos\theta \sin(\phi - \phi') + \hat{a}_\phi I_\phi \cos(\phi - \phi') \qquad (117)$$

where the source coordinates have been primed and the observation coordinates have not been primed. Since I_ϕ is a constant, by circular symmetry A_ϕ will not be a function of ϕ. Hence for simplicity we can select $\phi = 0$. Substituting the distance in terms of a spherical coordinate system, the potential function in eqn (115) could be written as

$$A_\phi = \frac{a\mu I_0}{4\pi} \int_0^{2\pi} I_\phi \cos\phi' \frac{e^{-jk\sqrt{r^2 + a^2 - 2ar\sin\theta\cos\phi'}}}{\sqrt{r^2 + a^2 - 2ar\sin\theta\cos\phi'}} \, d\phi' . \qquad (118)$$

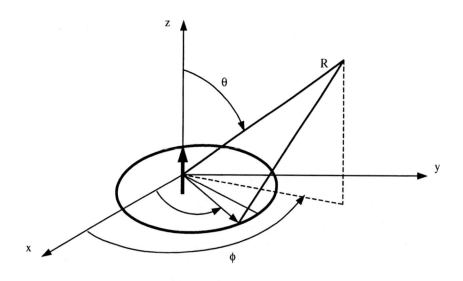

Figure 2.14: Small circular current loop.

It is not possible to perform the integration in eqn (118). Hence the integrand is approximated by the first two terms of the MaClaurin expansion about $a = 0$ to yield

$$A_\phi \approx \frac{a\mu I_0}{4\pi} \int_0^{2\pi} \cos\phi' \left[\frac{1}{r} + a\left(\frac{jk}{r} + \frac{1}{r^2} \right) \sin\theta\cos\phi' \right] e^{-jkr} \, d\phi' . \qquad (119)$$

So

$$A_\phi \approx \frac{a^2 \mu I_0}{4} e^{-jkr} \left(\frac{jk}{r} + \frac{1}{r^2} \right) \sin\theta . \tag{120}$$

By a similar analysis the r-component and the θ-component can be written as

$$A_r \approx \frac{a\mu I_0}{4\pi} \sin\theta \int_0^{2\pi} \sin\phi' \left[\frac{1}{r} + a\left(\frac{jk}{r} + \frac{1}{r^2} \right) \sin\theta \cos\phi' \right] e^{-jkr} \, d\phi', \tag{121}$$

$$A_\theta \approx \frac{a\mu I_0}{4\pi} \sin\theta \int_0^{2\pi} \sin\phi' \left[\frac{1}{r} + a\left(\frac{jk}{r} + \frac{1}{r^2} \right) \sin\theta \cos\phi' \right] e^{-jkr} \, d\phi', \tag{122}$$

but the integrations in eqns (121) and (122) yield nulls. Substituting eqn (120) into eqn (193) yields the components of the magnetic field $\mathbf{H}$ as

$$H_r = j\frac{ka^2 I_0 \cos\theta}{2r^2} \left[1 + \frac{1}{jkr} \right] e^{-jkr}, \tag{123}$$

$$H_\theta = -\frac{(ka)^2 I_0 \sin\theta}{4r} \left[1 + \frac{1}{jkr} - \frac{1}{(kr)^2} \right] e^{-jkr}, \tag{124}$$

$$H_\phi = 0 . \tag{125}$$

Using eqn (1) with $\mathbf{J} = 0$, the electric field components could be found as

$$E_r = E_\theta = 0, \tag{126}$$

$$E_\phi = \eta \frac{(ka)^2 I_0 \sin\theta}{4r} \left[1 + \frac{1}{jkr} \right] e^{-jkr}, \tag{127}$$

where η is the intrinsic impedance of free space.

2.10.3 *Field equations of a magnetic dipole*

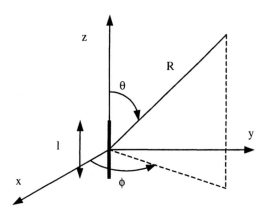

Figure 2.15: Magnetic dipole of length *l*.

The electric vector potential **F** defined by eqn (114) could be rewritten as

$$F = \frac{\varepsilon}{4\pi} \int_l I_m \frac{e^{-jkr}}{r} \, dz \,. \tag{128}$$

When considering current flow in an infinitesimal length l the electric vector potential reduces to

$$\mathbf{F} = \hat{a}_z \frac{\varepsilon I_m l}{r} e^{-jkr} \,. \tag{129}$$

Substituting eqn (129) into eqn (105) we obtain the electric field components as

$$E_r = E_\theta = 0 \,, \tag{130}$$

$$E_\phi = -j \frac{kI_m l \sin\theta}{4\pi r} \left[1 + \frac{1}{jkr} \right] e^{-jkr} \,. \tag{131}$$

The magnetic field **H** obtained using eqn (2) is given by

$$H_r = \frac{I_m l \cos\theta}{2\pi\eta r^2} \left[1 + \frac{1}{jkr} \right] e^{-jkr} \,, \tag{132}$$

$$H_\theta = j \frac{kI_m l \sin\theta}{4\pi\eta r} \left[1 + \frac{1}{jkr} - \frac{1}{(kr)^2} \right] e^{-jkr} \,. \tag{133}$$

Comparing eqns (131)–(133) with eqns (123)–(127), it can be seen that a small current loop is equivalent to an elemental magnetic dipole with its axis perpendicular to the plane of the current loop (see Fig. 2.15). The fields radiated by a magnetic dipole characterize the radiation fields of a hydrogen nucleus stimulated by a radio frequency pulse in MRI.

2.10.4   Magnetic field pattern of a magnetic dipole

Field patterns of a magnetic dipole in a plane vertical to the xy plane at a distance of 1 m (length of dipole $\lambda/50$, frequency = 100 MHz) are considered.

The magnetic field equations and the associated field patterns for a magnetic dipole are equivalent to those of the electric field of an electric dipole. The polar field patterns of the magnetic field components H_r and H_θ are shown in Fig. 2.16(a) and (b), respectively. These radiation patterns will be used in the analysis of the resonating nuclear magnets.

2.10.5   The helical broadband antenna

A helical antenna, shown in Fig. 2.17, is a conductor that is wound into a helical shape and is fed properly. It may be thought of as a combination, or an array, of small loop antennas.

The symbols used to describe the helix are defined as follows:

D = diameter of helix (between centers of coil material);
C = circumference of helix = πD;
S = spacing between turns;
α = pitch angle;
L = length of one turn;
N = number of turns;
A = axial length;
λ = wavelength.

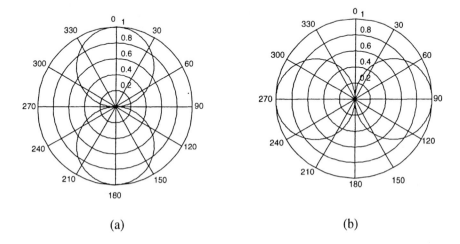

(a)   (b)

Figure 2.16: Normalized field pattern of (a) H_r and (b) H_θ on the vertical plane.

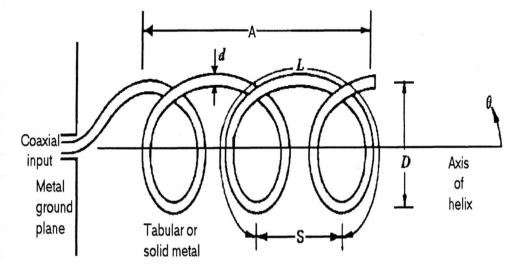

Figure 2.17: Geometry and dimensions of a helical antenna.

An infinitely long helix is a transmission line, which can support an infinite number of modes. Corresponding to these modes, a finite length helix can radiate in a number of modes. Two of these modes are the normal mode and the axial mode. The normal mode yields radiation that is at right angles to the axis of the helix. This occurs when the helix diameter is much smaller than one wavelength. The axial mode provides maximum radiation along the axis of the helix. This will occur when the helix circumference is one wavelength. Although the helical antenna radiates in different modes depending on conditions, the axial mode is often of most importance. The axial mode helical antenna is used in such diverse applications as printed circuit board probing and in satellite reception. The radiation patterns of the two different modes are as shown in Fig. 2.18.

If one turn of the helix is uncoiled, the relationships among the various helix parameters are as shown in Fig. 2.19. In the axial mode of radiation, the maximum radiation is along the axis of the helix, that is, the helix radiates as an endfire antenna. The axial mode occurs when the helix circumference is of the order of one wavelength. The axial mode carries a nearly pure traveling wave; therefore, the effect of the ground plane may be neglected. Also, the size of the ground plane is not very critical but it should be made wider than half a wavelength. The conductor diameter d has very little effect on the axial mode helix antenna properties, so a nominal value d may be chosen. The helix may be fed using a coaxial cable with the center conductor attached to the helix and the outer conductor attached to the ground plane that is a solid metal (copper) square plane. However it is not practical to implement a copper plate larger than half a wavelength, which is approximately 30 cm for a 500 MHz helical antenna.

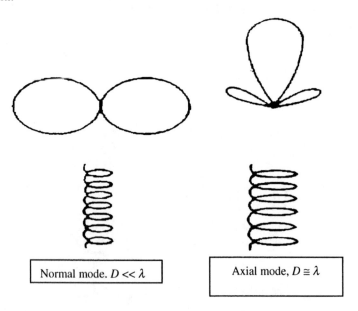

Normal mode. $D \ll \lambda$

Axial mode, $D \cong \lambda$

Figure 2.18: Radiation pattern of the two different modes.

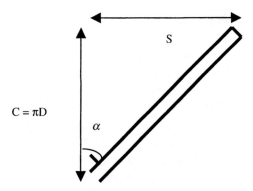

Figure 2.19: One uncoiled turn of a helix.

The dimensions of the helical antenna are worked out below. The required bandwidth BW = 300 MHz. The center or carrier frequency f_c = 250 MHz.

$$\lambda = \frac{c}{f_c},$$

$$\lambda = \frac{3 \times 10^8}{250 \times 10^6}$$

$$= 1.2 \text{ m.}$$

Range of C: $3/4\lambda \le C \le 4/3\lambda$.

The empirical formula for the half-power beam width (HPBW) of a helical antenna is as follows:

$$\text{HPBW} = \frac{52}{(C/\lambda)\sqrt{N(S/\lambda)}}. \qquad (134)$$

If we choose a typical configuration of $C = \lambda$, $\alpha = 12°$ and $N = 12$, then

$$S = C \tan \alpha,$$

a formula which holds for $12° \le \alpha \le 15°$, $N > 3$, and $3/4\lambda < C < 4/3\lambda$.

We let $N = 12$, $\alpha = 12° \Rightarrow S = 1.2 \tan 12° = 0.255$ m. Hence

$$\text{HPBW} = \frac{52}{(C/\lambda)\sqrt{N(S/\lambda)}} = 32.6°.$$

An empirical formula for the input resistance of a helical antenna is

$$R_{in} = 140(C/\lambda) \ \Omega, \qquad (135)$$

which yields an input impedance of 140 Ω for the 250 MHz helical antenna.

2.11   Effect of ground on antenna radiated electric fields

So far in this chapter all the analysis and discussions have assumed that the antenna is placed in free space and that it is far away from the ground, as shown in Fig. 2.20. Although the fields near elevated microwave antennas may closely approximate this idealized situation, the fields of most antennas are affected by the presence of the ground. The reason for this is that at microwave frequencies the antenna radiation beam is quite narrow, and the radiated wave energy will therefore not impinge on the ground if the antenna is well above the ground and pointed away from the ground. This is not the case for antennas with wide or isotropic beams. Any electromagnetic energy from the radiating element directed toward the ground undergoes a reflection (Fig. 2.21). The amount of reflected energy and its direction in which the reflected wave travels are determined by the geometry and constitutive parameters (i.e. conductivity and permittivity) of the ground. Since the reflected wave will now interfere with the main or direct wave from the antenna, there the overall radiation pattern of the antenna will be different from the free space radiation pattern of the antenna. Furthermore, the reflected energy will be frequency dependent since the effective conductivity of the ground (i.e. $\sigma + j\omega\varepsilon$) increases with frequency.

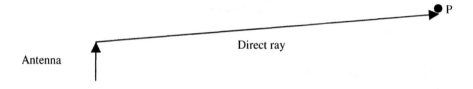

Figure 2.20: Total field strength at point P is only due to the direct ray from the antenna placed in free space.

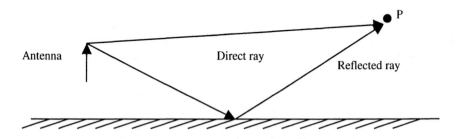

Figure 2.21: The direct ray and the reflected ray from the ground.

To analyze the situation where the reflection at the ground must be taken into account, we may replace the ground by the image of the antenna, as shown in Fig. 2.22. It is seen from Fig. 2.22 that the reflected wave (or ray) and the image ray have the same path lengths (i.e. AD = BD). At any observation point P, the resultant electromagnetic field strength will be the vector sum of the field strengths of the direct ray and the reflected ray. The reflected ray travels an extra distance of BC further than the direct ray, and thus at point P it would have a phase different from that of the direct ray. Although the amplitudes of the two rays will differ as well, we ignore this difference by assuming that AD is much greater than BC. It is the phase difference between the two rays that gives rise to signal cancelation or fading as in mobile communication systems. The application of the principle of images to find the resultant field strength at any point P applies not only to a perfectly conducting ground but also to ground of any conductivity. We will use this image principle to find the total field strength $\mathbf{E}_\theta$ and $\mathbf{E}_R$ components of a vertical and horizontal dipole transmitter at a distance D m from the receiver; the heights of the transmitting and receiving antennas from the ground are H_1 m and H_2 m, respectively.

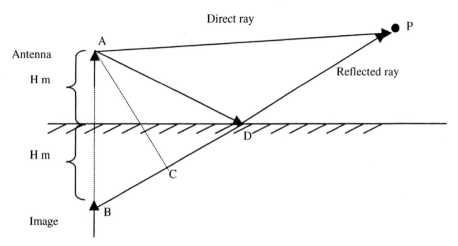

Figure 2.22: Illustration of the use of image concept to find total field strength at point P.

2.11.1   The vertical dipole

Figure 2.23 shows a vertical dipole antenna of length L and carrying a current I placed above a ground plane. The dipole antenna is assumed to be an infinitesimal dipole antenna. Figure 2.24 shows the electric field components of the direct and reflected rays.

Observing the situation in Fig 2.24, the direct ray component of the electric field $\mathbf{E}_\theta$ at the receiver is given as

$$\mathbf{E}_{\theta d} = j\eta_1 \frac{kIL \sin \theta_d}{4\pi} \left[\frac{1}{r_1} + \frac{1}{jkr_1^2} - \frac{1}{k^2 r_1^3} \right] e^{-jkr_1} \, \mathbf{u}_\theta, \qquad (136)$$

$$\mathbf{E}_{Rd} = \eta_1 \frac{IL \cos \theta_d}{2\pi} \left[\frac{1}{r_1^2} + \frac{1}{jkr_1^3} \right] e^{-jkr_1} \, \mathbf{u}_r . \qquad (137)$$

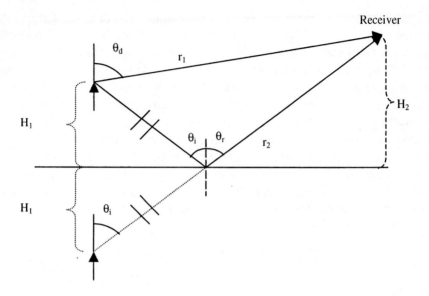

Figure 2.23: Multipath effects for a vertical dipole antenna.

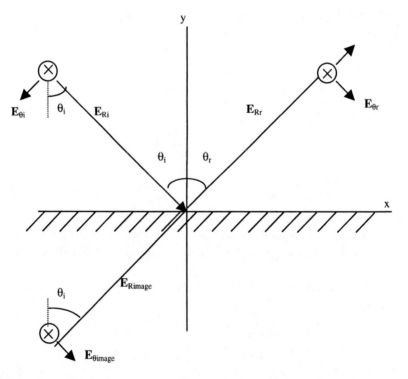

Figure 2.24: Reflected E field components $\mathbf{E}_\theta$ and $\mathbf{E}_R$ for a vertical dipole above a ground plane with finite conductivity.

The $\mathbf{E}_\theta$ and $\mathbf{E}_R$ components of the image shall be resolved into their x- and y-components in order to be able to get the resultant components by adding them to the fields of the direct ray. Thus we seek to obtain the $\mathbf{E}_x$ and $\mathbf{E}_y$ fields from the image. Subsequently we shall get the resultant $\mathbf{E}_x$ and $\mathbf{E}_y$ components at the observation point (i.e. the receiver). Once we know the x- and y-components, we can get the reflected $\mathbf{E}_\theta$ and $\mathbf{E}_R$ through these $\mathbf{E}_x$ and $\mathbf{E}_y$ components. From Fig. 2.24, for the electric fields at the receiver we get

$$\mathbf{E}_{xr} = \mathbf{u}_x(E_{Rimage} \sin \theta_i + E_{\theta image} \cos \theta_i), \tag{138}$$
$$\mathbf{E}_{yr} = \mathbf{u}_y(E_{Rimage} \cos \theta_i - E_{\theta image} \sin \theta_i), \tag{139}$$

where E_{Rimage} and $E_{\theta image}$ are given by eqns (137) and (136), respectively, by replacing r_1 by r_2 and θ_d by θ_i.

The reflection coefficients are given by

$$\Gamma_v = 1, \qquad \Gamma_h = \frac{\eta_2 \cos \theta_i - \eta_1 \cos \theta_t}{\eta_2 \cos \theta_i + \eta_1 \cos \theta_t}, \tag{140}$$

where

$\eta_1 = 120\pi$, the intrinsic impedance of free space,
$\eta_2 = [j\omega\mu/(\sigma + j\omega\varepsilon)]^{1/2}$, the intrinsic impedance of ground.

The angles of θ_i and θ_t are related by Snell's law of refraction:

$$n_1 \sin \theta_i = n_2 \sin \theta_t,$$

where

$$n_1 = \frac{c(\text{speed of light} = 3\times10^8)}{U_{p1}(\text{speed of ray in air medium})},$$

$$n_2 = \frac{c(\text{speed of light} = 3\times10^8)}{U_{p2}(\text{speed of ray in ground medium})}.$$

In most cases $n_1 = 1$ and $n_2 > 1$.

From Fig. 2.24, the reflected $\mathbf{E}_x$ and $\mathbf{E}_y$ components are given by

$$\mathbf{E}_{xr} = \mathbf{E}_{Rr} \sin \theta_r + \mathbf{E}_{\theta r} \cos \theta_r, \tag{141}$$
$$\mathbf{E}_{yr} = \mathbf{E}_{Rr} \cos \theta_r - \mathbf{E}_{\theta r} \sin \theta_r. \tag{142}$$

According to Snell's law of reflection, $\theta_i = \theta_r$. By solving eqns (141) and (142), the reflected $\mathbf{E}_\theta$ and $\mathbf{E}_R$ components are given by

$$\mathbf{E}_{\theta r} = \mathbf{E}_{xr} \cos \theta_r - \mathbf{E}_{yr} \sin \theta_r, \tag{143}$$
$$\mathbf{E}_{Rr} = \mathbf{E}_{xr} \sin \theta_r + \mathbf{E}_{yr} \cos \theta_r. \tag{144}$$

Thus, considering both the direct ray and the reflected ray, the total fields for each individual component at the receiver are given by

$$\mathbf{E}_{\theta total} = \mathbf{E}_{\theta r} + \mathbf{E}_{\theta d},$$ (145)
$$\mathbf{E}_{Rtotal} = \mathbf{E}_{Rr} + \mathbf{E}_{Rd}.$$ (146)

Therefore the resultant amplitude of the total field at the receiver is given by

$$E_{total} = \sqrt{|\mathbf{E}_{\theta total}|^2 + |\mathbf{E}_{Rtotal}|^2}\ .$$ (147)

Alternatively, we can express the fields at the receiver in terms of x- and y-components as follows:

$$\mathbf{E}_{xtotal} = \mathbf{E}_{xr} + \mathbf{E}_{xd},$$ (148)
$$\mathbf{E}_{ytotal} = \mathbf{E}_{yr} + \mathbf{E}_{yd},$$ (149)

where

$\mathbf{E}_{xd}$ = direct ray x-component at the receiver
$\quad = \mathbf{E}_R \sin\theta_d + \mathbf{E}_\theta \cos\theta_d,$ (150)
$\mathbf{E}_{yd}$ = direct ray y-component at the receiver
$\quad = \mathbf{E}_R \cos\theta_d - \mathbf{E}_\theta \sin\theta_d.$ (151)

2.11.2 The horizontal dipole

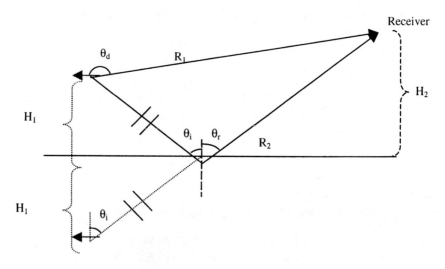

Figure 2.25: Horizontal dipole above a ground plane.

The formulae derived for the vertical dipole can be used for the horizontal dipole (see Fig. 2.25) to get the resultant signal at the receiver except that the θ_i used in the formula for $\mathbf{E}_{\theta imageR2}$ and $\mathbf{E}_{rimageR2}$ need to be replaced by $\pi/2 + \theta_i$ as shown below:

$$\mathbf{E}_{\theta\text{image}} = j\eta_1 \frac{kIL\sin(\pi/2+\theta_i)}{4\pi}\left[\frac{1}{r_2}+\frac{1}{jkr_2^2}-\frac{1}{k^2r_2^3}\right]e^{-jkr_2}\,\mathbf{u}_\theta,$$

$$\mathbf{E}_{R\text{image}} = \eta_1 \frac{IL\cos(\pi/2+\theta_i)}{2\pi}\left[\frac{1}{r_2^2}+\frac{1}{jkr_2^3}\right]e^{-jkr_2}\,\mathbf{u}_r.$$

The direct ray field components are given by

$$\mathbf{E}_{xd} = -(\mathbf{E}_R\cos\theta_d + \mathbf{E}_\theta\sin\theta_d), \tag{152}$$
$$\mathbf{E}_{yd} = \mathbf{E}_R\sin\theta_d + \mathbf{E}_\theta\cos\theta_d. \tag{153}$$

Simulation results for vertical and horizontal dipole as the distance between transmitter and receiver varies are given in Figs 2.26 and 2.27. The transmitter and receiver are both kept at a height of 78 m, as in the case of a typical airport air-traffic control tower.

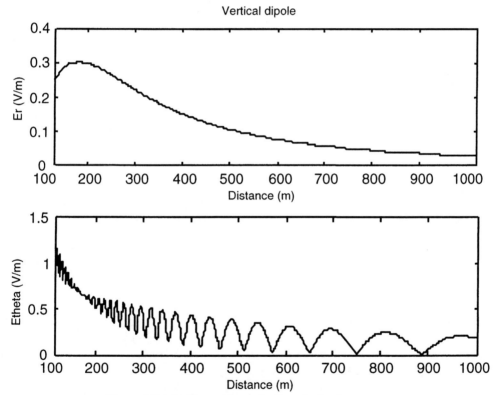

Figure 2.26: Fading with vertically polarized antenna.

The plots show $\mathbf{E}_\theta$, $\mathbf{E}_R$, $\mathbf{E}_{\text{total}}$, $\mathbf{E}_x$, $\mathbf{E}_y$ components at the receiver for a signal frequency of 120 MHz. The distance between the transmitter and the receiver varies from 100 to 300 m. The frequency of 120 MHz was chosen since it is one of the 760 channel frequencies used for air-traffic control. The conductivity of ground was assumed to be $\sigma = 99999999 \approx \alpha$, the conductivity of a perfect conductor ground. The transmitter current is 1 A, and the relative permittivity of the ground $\varepsilon_R = 25$. Figure 2.26 shows the results for a vertically polarized antenna, and Fig. 2.27 shows the results for a horizontally polarized antenna. The vertically polarized antenna demonstrates rapid fading of signal with distance, known as Rayleigh fading due to the reflected signal destructively interfering with the direct signal.

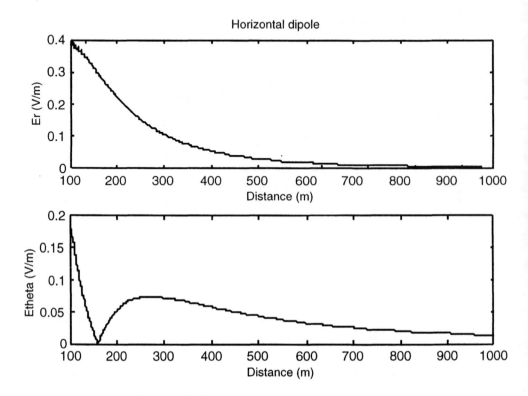

Figure 2.27: Fading with horizontally polarized antenna.

2.12   Frequency independent antennas

In certain special applications antennas that may be used over very wide frequency bandwidth may be required. This means that the input impedance and radiation pattern of the antenna must be constant over a large frequency range. An example of such an antenna is the Vivaldi antenna shown in Fig. 2.28.

The curved guiding structure is defined by $y = A \exp(kx)$. Waves travel along the curved structure and are radiated out. When the spacing between the two curved conductors is small, very little radiating takes place, and the signals cling to the conductors as they travel. However as they near the end of the curved structure, the distance between conductors is large, and the electromagnetic energy is released to be radiated away. The gain of the Vivaldi antenna is due to the signal phase velocity inside the curved structure being equal or greater than that in the medium outside the antenna. When the outside medium is free space, the phase velocity of the signal is 3 × 10^8 m/s, and the velocity inside the antenna is greater than or equal to this. Since high frequency signals have a higher phase velocity than a low frequency signal, an impulse applied to the frequency independent antenna will appear as a linear-frequency modulated (FM) signal (i.e. a chirp signal) at the output of the antenna. The higher frequency signals appear first at the output with the lowest frequency component appearing last at the output, and hence the chirp waveform at the output of the antenna.

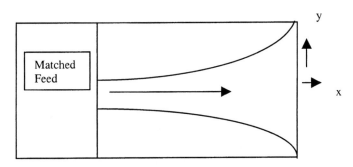

Figure 2.28: The Vivaldi antenna.

Other forms of frequency independent antennas are the biconical dipole, equiangular spiral, conical spiral and log-periodic structures. In general, to obtain such a frequency independent antenna its geometrical features are entirely defined by angles. In such an antenna its physical surface must satisfy the equation $r = f(\theta) \exp(k\phi)$ in the spherical coordinate system (r,θ,ϕ). Absorbing material may be placed on one side to get unidirectional radiation with a planar equiangular spiral antenna, which radiates circularly polarized waves.

Acknowledgement

Ng Joo Seng and Seow Chee Kiat developed the MATLAB™ programs contained on disk (Program 2.1). Sections 2.9 and 2.10 are based on Naveendra (1999) and Section 2.11 on Ng and Seow (1998).

3 Focused beam antennas

P.R.P. Hoole

3.1 Introduction

The wire antennas we looked at in the last chapter are largely antennas with wide beams; the half-power beam width (HPBW) is of the order of a few tens of degrees, but in many applications, like line of sight microwave communications, satellite communications and radar, the antenna should have an HPBW of the order of 0.2–1°. In such cases we need narrow beam antennas. Two widely used narrow beam antennas are the array antennas, which also permit electronic control of the beam direction (beamsteering), and the aperture antennas.

In an array antenna a number of radiating elements are arranged together to improve power gain and directivity. The individual antenna elements may be arranged along a straight line (linear array) or in a square grid (planar array) (see Fig. 3.1). Other forms of arrangements are also possible; in radio astronomy, for instance, large parabolic reflector type aperture antennas are arranged in a star formation.

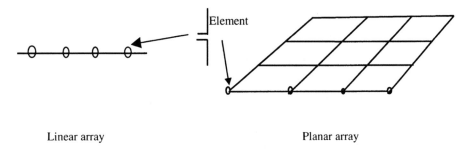

Linear array   Planar array

Figure 3.1: Array antennas.

Consider an array antenna made up of nine wire antennas. Each element is connected to a source voltage of $V_S = \text{Re}(Ve^{j\omega t})$; the current flowing in each wire antenna element is $I = V/Z$. Then the resultant radiated electric field $E_T = 9E$, where E is the electric field radiated by each wire antenna. The resultant power density of the radiated signal is $\frac{1}{2}(9E)^2/120\pi = 40.5E^2/120\pi$ W/m^2. Here $\eta = 120\pi$ Ω is the intrinsic or wave impedance of free space. If there were only one wire antenna, then the radiated power density would have been only $0.5E^2/120\pi$ W/m^2. By using an array of nine antennas (making up an array antenna) instead of a single antenna, we have increased the radiation power by (40.5/0.5), i.e. 81 times! Similarly, if the same array antenna were used to receive signals, its reception efficiency will be 81 times better than that of a single antenna element. Two further advantages of the phased array antenna are (i) the antenna beam can be electronically steered without any need for a mechanical or inertial system to turn the antenna and (ii) the array antenna can be placed on airframes without altering the aerodynamics of the vehicle.

Two requirements which should be met when designing array antennas are (i) fields from separate elements must be in phase at the receiving antenna, and (ii) currents in all elements must be identical in magnitude. Unless these requirements are met, for instance if the phases of the received signals are not in phase at each element, then there may be destructive interference at the receiver. When array antennas are used for transmission they may have two types of elements: (a) driven elements (these are antenna elements to which the signal source V_S is connected); and (b) parasitic elements (these are antenna elements to which the source is not directly connected). They receive power from the driven elements through coupling. Some parasitic elements are placed in front of driven elements to narrow the radiation beam, and are called directors. Other parasitic elements are placed behind the driven element to prevent signals going in the backward direction and to focus all radiation energy in one single forward direction. These parasitic elements are called reflectors. The large parabolic dishes placed behind the horn or wire antennas in microwave antennas are reflectors. The driven element–parasitic element combination array antennas are also used as receiver antennas.

3.2   Array antennas: two-element linear array

An array antenna is a system comprising a number of radiating elements, generally similar. A linear array antenna is a number of antenna elements placed in a straight line. It yields a narrow, steerable fan beam. To get a narrow, steerable pencil beam we need to have a planar array in which, for instance, the antenna elements should be arranged on a square grid. Mechanically steerable 100×50 cm fixed-pencil beam array antennas are used in most fighter aircraft, usually mounted on the nose of the aircraft and costing about US$ 2 million. In this section we shall consider a simple case, one in which the radiators are all identical, all oriented in the z direction, all aligned on the y-axis, and all equally spaced. The z-directed dipoles are placed on the x-axis, with an interelement separation of d. We shall see that when similar antennas are arranged in various configurations with proper amplitude and phase relations, we obtain desired radiation characteristics, e.g. direction and width of main beam, sidelobe levels, directivity.

3.2.1   Two-element Hertzian dipole array antenna

We shall consider briefly two infinitesimal (Hertzian) dipoles used to form an array antenna. The currents flowing into the two elements are out of phase by an electrical angle δ. The phase shift may be obtained by using both elements from a single source, so that the current magnitude is I for both antenna elements, but by installing a phase shifter in the cable run between the first and the second element, the current supplied to the second element may be phase shifted by any desired angle δ (see Fig. 3.2).

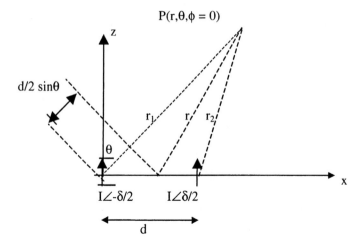

Figure 3.2: Two-element array with the observation point on the zx-plane.

We shall show that a microprocessor, for instance, controlling the steps of phase shift δ can steer the radiation pattern or beam of the antenna in any required direction. A simpler method will be to place a capacitor in series along the line, or a microstrip line of length $\lambda/4$, between the two elements to give a phase shift of 90°. The distance between the two elements is d. For convenience we assume that the observation point is on the zx-plane ($\phi = 0$), and that it subtends an angle θ with the z-axis. By allowing the observation point $P(r,\theta,\phi = 0)$ to be on the zx-plane, we look for a resultant radiation field which will not be a function of ϕ, thus making the initial analysis simple; later, when we consider the radiation field of a two-element array made of half-wave dipoles, we shall allow the observation point to be at $P(r, \theta, \phi)$.

Furthermore, the observation point is assumed to be in the far-field zone of the antenna, so that only the $E(1/r)$ and $H(1/r)$ components exist. Hence the electric field we shall be interested in is the $\mathbf{u}_\theta$ component; the vector sign will be dropped in the following analysis, with the understanding that the electric field we obtain is the E_θ component.

Using radiation field for an infinitesimal wire antenna, eqn (47) in Chapter 2 in particular,

$$E = \mathbf{u}_\theta \, j\eta \frac{kIL}{4\pi r_1} e^{-jkr} \sin\theta, \tag{1}$$

and we may write the electric field intensities radiated by the two elements as follows:

$$E_1 = \mathbf{u}_\theta \, j\eta \frac{kIL}{4\pi r_1} e^{-j(kr_1 + \delta/2)} \sin\theta_1, \tag{2}$$

$$E_2 = \mathbf{u}_\theta \, j\eta \frac{kIL}{4\pi r_2} e^{-j(kr_2 - \delta/2)} \sin\theta_2. \tag{3}$$

Hence the resultant field at the observation point P is given by

$$E_p = E_1 + E_2$$
$$= u_\theta j\eta \frac{kIL}{4\pi} \left[\frac{e^{-j(kr_1+\delta/2)} \sin\theta_1}{r_1} + \frac{e^{-j(kr_2-\delta/2)} \sin\theta_2}{r_2} \right], \tag{4}$$

where we have assumed that $r_1, r_2 \gg d$ and $\theta_1 \approx \theta_2 \approx \theta$.

Amplitude: As far as the amplitude of the resultant field is concerned, we may assume that the three distances are equal, since the differences which appear will make very little difference when r is very large (far-field zone). Hence in the denominator, we let $r_1 = r_2 = r$.

Phase: As far as the phase is concerned, for a high frequency signal, even a small difference in distance in the exponential term will make a large difference in the phase of the resultant signal. The phases of the electric fields radiated by the two elements may interact destructively or constructively to significantly decrease or increase the resultant field magnitude. Indeed we could get total signal fading if the two signals are out of phase by π radians at the observation point. Hence for the phase term appearing in the exponential, we set

$$r_1 = r + d/2 \sin\theta \quad \text{and} \quad r_2 = r - d/2 \sin\theta.$$

Thus the resultant electric field is given by

$$E_p = u_\theta j \frac{\eta kILe^{-jkr}}{4\pi r} \sin\theta \left[e^{-j(kd \sin\theta+\delta)/2} + e^{j(kd \sin\theta+\delta)/2} \right]. \tag{5}$$

Using $e^{jx} = \cos x + j \sin x$, we may express the resultant field for an observation point on the zx-plane ($\phi = 0$) as

$$E_p = u_\theta j \frac{\eta kILe^{-jkr}}{4\pi r} \sin\theta \left[2\cos\left(\frac{kd \sin\theta+\delta}{2} \right) \right] \tag{6}$$

$$= \text{single element pattern} \times [\text{AF}], \tag{7}$$

where the array factor AF is given by

$$\text{AF} = 2\cos\left(\frac{kd \sin\theta+\delta}{2} \right). \tag{8}$$

The array factor AF is largely dependent on d (the distance between the elements) and δ (the phase angle difference between the electric currents in the two elements). For linear array antennas, the array factor is the same in the E-plane and in the H-plane. (For the geometry we are discussing, the E-plane is the zx-plane and the H-plane is the xy-plane.) The array factor is symmetrical about the axis on which the array of elements is placed. When the radiation pattern is to be sketched by hand it is common to sketch the H-plane pattern, since in the H-plane the radiation pattern is identical to the array factor. In the E-plane, the radiation pattern is the array factor multiplied by the sin θ factor that appears in the E-plane elemental pattern; this is more difficult to sketch.

In order to illustrate the generation of the AF pattern, we could try sketching the radiation patterns, and compare with the computer generated patterns shown in Fig. 3.3, for a two-element antenna with the following antenna parameters:

1. $d = \lambda/4$, $kd = \pi/2$, $\delta = \pi$, AF $= 2\sin(\pi/4\sin\theta)$,
2. $d = \lambda/2$, $kd = \pi$, $\delta = 0$, AF $= 2\cos(\pi/2\sin\theta)$,
3. $d = \lambda/4$, $kd = \pi/2$, $\delta = \pi/2$, AF $= 2\cos(\pi/4(\sin\theta + 1))$.

The plots shown in Fig. 3.3 are computer generated for the following parameters: $f = 900$ MHz, dipole length $L = 0.5\lambda$, interelement spacing $d = 0.5\lambda$, distance $r = 100$ m and phase δ is changed. The MATLAB^{TM} listing of the computer program used to generate the radiation patterns is also given in the text.

The listing of the program used for plotting the radiation field patterns follows:

```
% E field pattern
% MATLAB program by T. Naveendra
th = pi/100 : pi/100 : 2*pi ;
ph = pi/100 : pi/100 : 2*pi ;
et = 120*pi ; f = 100e6 ; c = 3e8 ;
n = input('Enter Antenna Length as Fraction of wavelength :') ;
k = 2*pi*f/c ; l = 3e8/(n*f) ; i = 10 ;
r = 1000 ; % meters
q = et*i/(2*pi*r) ;
s = k*l ;

for i = 1 : 200
  for j = 1 : 200
    e(i,j) = q*(cos(s*cos(th(i)))-cos(s))/sin(th(i)) ; ;
    x(i,j) = abs(e(i,j))*sin(th(i))*cos(ph(j)) ;
    y(i,j) = abs(e(i,j))*sin(th(i))*sin(ph(j)) ;
    z(i,j) = abs(e(i,j))*cos(th(i)) ;

  end
end

surf(x,y,z) ;
```

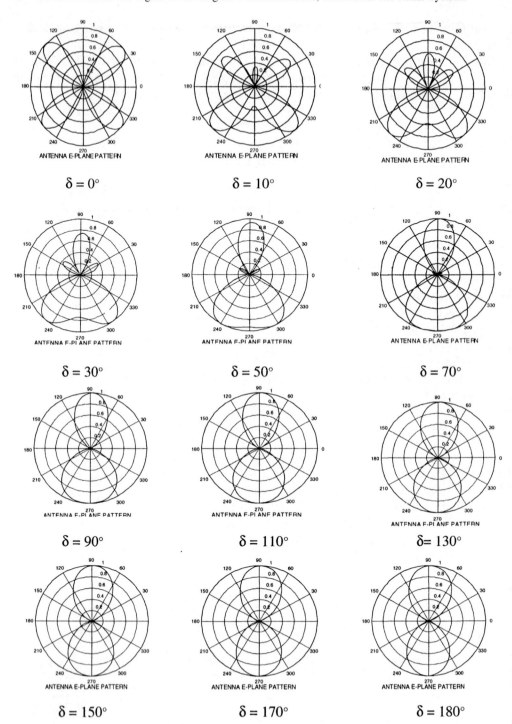

Figure 3.3: Radiation patterns of a two-element array.

From the resultant field patterns we observe two general classes of beams emerging. These are (a) the *endfire array* and (b) the *broadside array*. In the endfire array, the major part of the radiation pattern is directed along the axis y, which is the axis on which the elements are arranged. There is very little radiation in the z and x directions. In the broadside array, the radiation pattern is directed at right angles to the y-axis, the axis along which the elements are arranged. The radiation is in the z direction, and there is very little radiation in the x and y directions. An endfire array will be useful, for instance, when transmission may be required only in one direction, as in the case of communication systems to transmit to ships in the sea lane. Such antennas should avoid transmitting back into the land area. The endfire array is used for television reception as well; the Yagi–Uda antennas mounted on most rooftops are endfire arrays with the beam pointed towards the television broadcasting tower antenna. Indeed, such an antenna is attractive for mobile telephones, where it is desirable to radiate signals only away towards the base station and not into the head of the user. Here again a directive antenna like the endfire array is useful.

Another important feature about array antennas, which form the radiation patterns for changing phase angle δ, is that the direction of the beam may be controlled by changing the electrical phase difference δ between the two elements. This means that the array antenna beam may be electronically steered. Without using any mechanical device like a motor to direct the antenna in a particular direction, by changing δ we may make the antenna beam look in a particular direction. This is useful, for instance, in radar systems, where one may require the antenna to be continuously searching for targets in all directions. In mobile communications, smart antennas which exploit this feature may transmit power out to mobile users who may be moving around and are clustered in a certain area. A mobile base station which transmits power out in all directions (omni-directional antenna) is wasteful in power since the mobile stations may not be spread out over the whole area. Therefore a base station which tracks all the mobile users in its cell may use an electronically steerable antenna to focus its transmission on to the specific area in which the mobile users are located at any instant of time.

Phase shifters are low power modules placed at the input of the transmit amplifier (i.e. the output of the receive amplifier in a two-way communication array antenna). If the phase shifters are placed at the output of the transmit amplifier, there will need to be high power devices that are less accurate and slower. We shall later see that in adaptive arrays we not only control the phase (δ) of each element, but also the amplitude (I) of each element; control of amplitudes allows us to better control the sidelobes and null points of the radiation pattern. Analog type narrow-band phase shifters may use switched lines, where transmission lines of different lengths are switched into the path at the input of the amplifier. Switching is done by PIN diodes or FETs and are digitally controlled. Although these are easily implemented, losses of the order of 0.5 dB per section are present. If the number of line sections or bits is M, then the phase shifter resolution is $\Delta\theta = 2\pi/2^M$. For an M-bit phase shifter the usable fractional signal bandwidth is given by $\Delta f/f_c = 1/2^M$, where f_c is the carrier frequency. For the extreme case where the phase shift is 2π, the accuracy is $2\pi(\Delta f/f_c)$ rad. For broadband antennas, all-pass filter networks must be used instead of line lengths. These 2–5 bit phase shifters have an insertion loss of about 1 dB.

3.2.2   Two-element half-wave dipole array antenna

Consider an array of two half-dipole elements. Referring to Fig. 3.2, we set $r_1 = R_0$, the reference distance, and $r_2 = R_1$. The observation point P is at (r, θ, ϕ). For element 1, which is the reference element at the origin, the radiated electric field is given by

$$E_0 = E_m F(\theta, \phi) \frac{e^{-jkR_0}}{R_0},$$
(9)

and for element 2, which is at a distance d from the origin, the electric field is given by

$$E_1 = E_m F(\theta, \phi) \frac{e^{j\delta} e^{-jkR_1}}{R_1},$$
(10)

where $F(\theta, \phi)$ = pattern function of individual antennas = $\cos((\pi/2)\cos\theta)/\sin\theta$ for a half-dipole antenna. The pattern function of the individual elements will differ for E- and H-planes; in the case of the half-wave dipole we saw that it is a circle in the H-plane, whereas in the E-plane the elemental pattern function had regions of zero radiation along the axis on which the wire is placed. The array factor, on the other hand, is identical for both E- and H-planes. Hence it is the elemental radiation pattern (which depends on the type of antenna element used) which determines the difference in the overall radiation pattern in the E- and H-planes. Hand-drawn sketches of the radiation patterns for array antennas are mostly drawn for the H-plane since the elemental pattern is a circle for wire antenna elements.

The resultant electric field at the observation point P is, therefore,

$$E = E_m F(\theta, \phi) \left[\frac{e^{-j\beta R_0}}{R_0} + \frac{e^{j\delta} e^{-j\beta R_1}}{R_1} \right].$$
(11)

For the denominator, we may assume that the distances are equal, and for the phase term we may use the following approximation:

$$R_1 \approx R_0 - d \sin\theta \cos\phi.$$
(12)

Thus,

$$E = E_m \frac{F(\theta, \phi)}{R_0} e^{-jkR_0} \left[1 + e^{j(kd \sin\theta \cos\phi + \delta)} \right]$$
(13)

$$= E_m \frac{F(\theta, \phi)}{R_0} e^{-jkR_0} e^{j\psi/2} \left(2 \cos \frac{\psi}{2} \right),$$
(14)

where

$$\psi = kd \sin\theta \cos\phi + \delta.$$
(15)

We write the magnitude of the resultant field as

$$|E| = \frac{2E_m}{R_0} |F(\theta,\phi)| \left|\cos \psi/2\right|, \tag{16}$$

where $F(\theta,\phi)$ = element factor and $\left|\cos \psi/2\right|$ = normalized array factor.

Although the array factor and the normalized array factor will have similar patterns, the normalized array factor is always less than or equal to 1. Observe an important property of the normalized array factor: it is symmetrical about the axis of the array. In other words, with the antenna elements arranged along the x-axis, the array factor pattern is symmetrical about the x-axis. This may be verified by sketching the array factor for the zx-plane and the xy-plane. The array pattern on both planes will be identical, since in the zx-plane ($\phi = 0$), $\psi = kd \sin \theta + \delta$ and in the xy-plane ($\theta = \pi/2$), $\psi = kd \cos \phi + \delta$. Sketching polar plots of the array factor, or the normalized array factor, for these two ψ forms, we shall get the same array factor pattern. The elemental pattern $F(\theta,\phi)$ may not be symmetrical about the x-axis, and thus the resultant radiation pattern is normally not symmetrical about the x-axis. The elemental pattern $F(\theta,\phi)$ depends on the type of antenna elements used (e.g. dipole antenna elements, horn antenna elements, parabolic aperture antenna elements, etc.), but the array factor for a given d and δ is the same for all types of antenna elements, and it is thus independent of the type of antenna elements we use.

Therefore the pattern function of an array of identical elements is described by the product of the two directed half-dipoles:

$$|E| = \frac{2E_m}{R_0} \left|\frac{\cos\left((\pi/2)\cos \theta\right)}{\sin \theta}\right| \left|\cos \frac{\psi}{2}\right|, \tag{17}$$

where ψ is also a function of θ. The E-plane pattern (constant ϕ and θ varies) and the H-plane pattern (ϕ varies, $\theta = \pi/2$) are completely determined by $\cos \psi/2$. When we sketch the electric field radiation pattern of the resultant field, it is important to observe that the radiation pattern in the θ (or ϕ) and ψ coordinates resembles that of a low pass filter. In other words, electric field is peak for a certain value of θ (or ϕ, say $\phi = 0$, or $\phi = \pi/2$) and then it begins to fall off. Then there are sidelobes of much weaker strength. This output signal pattern very much resembles that of an output signal from a low pass filter, although in the low pass filter electronic circuit, filtering occurs in the time or frequency domains. Therefore an antenna may be considered as a spatial filter, that is, its output is a filtered version in the spherical space coordinates θ and ϕ. When we discuss the aperture antenna, we shall see that the radiation pattern is both θ and ϕ dependent. This observation is important in that, when we do antenna beamforming and synthesis, we shall use the same techniques as are used in electronic filter circuit synthesis to design our antenna system to provide a user specified radiation pattern or beam.

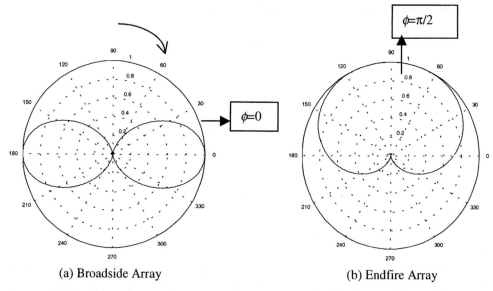

(a) Broadside Array　　　　　　(b) Endfire Array

Figure 3.4: Broadside and endfire beams with two-element and N-element arrays.

Example 1. Obtain the H-plane pattern of two parallel dipoles for (a) $d = \lambda/2$, $\delta = 0$ and (b) $d = \lambda/4$, $\delta = -\pi/2$. In the H-plane each dipole is omni-directional. Hence normalized pattern function = normalized array factor

$$\left|AF(\phi)\right| = \left|\cos\frac{\psi}{2}\right| = \left|\cos\left(\frac{1}{2}(kd\cos\phi + \delta)\right)\right|.$$

(a) $d = \lambda/2$, $kd = \pi$, $\delta = 0$ (Fig. 3.4(a)).

$$\left|AF(\phi)\right| = \left|\cos\left(\frac{\pi}{2}\cos\phi\right)\right|,$$

maximum at $\phi = \pm\,\pi/2$ in the broadside direction. It is possible to obtain this broadside beam with an N-element array, as we shall show in the next section.

(b) $d = \lambda/4$, $kd = \pi/2$, $\delta = -\pi/2$ (Fig. 3.4(b)).

$$\left|AF(\phi)\right| = \left|\cos\left(\frac{\pi}{4}(\cos\phi - 1)\right)\right|,$$

maximum at $\phi = 0$ resulting in an endfire array like the Yagi–Uda antenna.

3.3   General N-element uniform, linear array

We now consider an array of N antenna elements placed along the x-axis. The array is uniform in that the distance between adjacent elements is d and each element carries a current of I. The array is linear in that the phase angles of the currents in the elements

$n = 0, 1, 2, \ldots, (N-1)$ are $0, \delta, 2\delta, 3\delta, \ldots, (N-1)\delta$. Microstrip lines or microprocessors may provide these phase shifts. Antenna elements equipped with electronic phased shifters are called phased arrays. These can scan in θ (elevation) and ϕ (azimuth) directions. Widely used in radar and radio-astronomy systems, time delay circuits are used to provide phase shifts. By changing source frequency, we may change time delays. These are called frequency scanning arrays. N antenna elements are equally spaced and are excited with equal magnitude currents, with a phase displacement of $n\delta$ (see Fig. 3.5). We shall consider a radiation field pattern in the $\theta = \pi/2$ plane; hence the array factor is only a function of ϕ, and $\psi = kd \cos \phi + \delta$, since $\sin \theta = 1$.

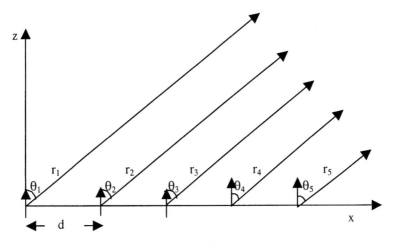

Figure 3.5: Uniform N-element linear array antenna.

At far field, the observation point P in Fig. 3.5 is very far from the antenna array and the static field and the induction field is much smaller than the radiating field. As a result, we could approximate the angles as

$$\theta_1 \approx \theta_2 \approx \theta_3 \approx \theta_4 \approx \theta_5 \approx \theta.$$

Further, to determine the amplitude of the electric field signal we could set

$$r_1 \approx r_2 \approx r_3 \approx r_4 \approx r_5 \approx r,$$

whereas the phase of the signal is allowed to have the following distance shifts with respect to the reference distance to allow for the important phase shifts between the elements:

$$r_2 = r - d \sin \theta \cos \phi,$$
$$r_3 = r - 2d \sin \theta \cos \phi,$$
$$r_4 = r - 3d \sin \theta \cos \phi,$$
$$r_5 = r - 4d \sin \theta \cos \phi.$$

In the following discussion, for the sake of simplicity, we shall set $\phi = 0$, or $\cos \phi = 1$. Assuming all the elements have the same current amplitude and dimension, the resultant field for an N-element antenna becomes

$$E_{\theta \text{ total at point } p} = E_{\theta 1} + E_{\theta 2} + E_{\theta 3} + E_{\theta 4} + \cdots + E_{\theta N}$$

$$= \frac{j \eta k I L \sin \theta \, e^{-jkr}}{4 \pi r} \left[1 + e^{j(kd \sin \theta + \delta)} + e^{j2(kd \sin \theta + \delta)} + \cdots + e^{j(N-1)(kd \sin \theta + \delta)} \right], (18)$$

$$E_{r \text{ total at point } p} = E_{r1} + E_{r2} + E_{r3} + E_{r4} + E_{r5}$$

$$= \frac{\eta I L \cos \theta \, e^{-jkr}}{2 \pi r^2} \left[1 + e^{j(kd \sin \theta + \delta)} + e^{j2(kd \sin \theta + \delta)} + \cdots + e^{j(N-1)(kd \sin \theta + \delta)} \right]$$

$$\approx 0. \tag{19}$$

It is apparent from eqns (18) and (19) that the total field of the array is equal to the field of a single element positioned at the origin multiplied by a factor which is widely referred to as the array factor. Thus, in our case, for a five-element array of constant amplitude, the array factor is given by

$$AF = 1 + e^{j(kd \sin \theta + \delta)} + e^{j2(kd \sin \theta + \delta)} + \cdots + e^{j(N-1)(kd \sin \theta + \delta)}. \tag{20}$$

The array factor is a function of the geometry of the array and the excitation phase. Thus by varying the separation d and/or the phase δ between elements, the characteristics of the array factor and of the total field of the array can be controlled. In general, we may state that the total electric field radiated by an array antenna is given by

$$E(\text{total}) = E(\text{single element}) \times \text{array factor (AF)}. \tag{21}$$

This is referred to as *pattern multiplication* for arrays of identical elements. Thus for an N-element array, the array factor can be written as

$$AF = 1 + e^{j(kd \sin \theta + \delta)} + e^{j2(kd \sin \theta + \delta)} + \cdots + e^{j(N-1)(kd \sin \theta + \delta)} \tag{22}$$

$$= \sum_{n=1}^{N} e^{j(n-1)(kd \sin \theta + \delta)} \tag{23}$$

$$= \sum_{n=1}^{N} e^{j(n-1)\psi}, \tag{24}$$

where $\psi = kd \sin \theta + \delta$.

In general, $\psi = kd \sin \theta \cos \phi + \delta$. By multiplying both sides of eqn (24) by $e^{j\psi}$, it can be written as

$$AF(e^{j\psi}) = e^{j\psi} + e^{j2\psi} + \cdots + e^{j(N-1)\psi} + e^{jN\psi}. \tag{25}$$

Subtracting eqn (22) from eqn (25) reduces to

$$AF(e^{j\psi} - 1) = (-1 + e^{jN\psi}), \tag{26}$$

which can be written as

$$AF = \frac{e^{jN\psi} - 1}{e^{j\psi} - 1} e^{j[(N-1)/2]\psi} \frac{e^{j(N/2)\psi} - e^{-j(N/2)\psi}}{e^{j(1/2)\psi} - e^{-j(1/2)\psi}}$$

$$= e^{j[(N-1)/2]\psi} \left[\frac{\sin(N/2)\psi}{\sin(1/2)\psi} \right]. \tag{27}$$

Expressing the foregoing analysis in another form, the total electric field E is the electric field E_0 due to a single element multiplied by the array factor. The total field is

$$|E_t| = |E_0| \left| 1 + e^{j\psi} + e^{j2\psi} + \cdots + e^{j(N-1)\psi} \right|$$

$$= |E_0| \left| \frac{1 - e^{jN\psi}}{1 - e^{j\psi}} \right| = |E_0| \left| \frac{e^{j(N/2)\psi}}{e^{j\psi/2}} \right| \frac{\sin N\psi/2}{\sin \psi/2}$$

$$= |E_0| \left| \frac{\sin N\psi/2}{\sin \psi/2} \right|, \tag{28}$$

where $\psi = kd \cos \phi + \delta$ and the magnitude of

$$\left| \frac{e^{j(N/2)\psi}}{e^{j\psi/2}} \right|$$

is equal to 1.

We define the normalized array factor

$$|AF(\psi)| = \frac{1}{N} \left| \frac{\sin N\psi/2}{\sin \psi/2} \right|. \tag{29}$$

The advantage in handling the normalized array factor instead of the array factor is that the normalized array factor is always less than or equal to 1, irrespective of the number of antenna elements making up the array antenna. Other than the difference in amplitudes, the patterns obtained for both the normalized array factor and the proper array factor are identical. The actual radiation pattern as a function of ϕ depends on kd ($=2\pi d/\lambda$) and δ. As $\phi : 0 \to 2\pi$ we have $\psi : kd + \delta \to -kd + \delta$ covering a range of $2kd$ ($=4\pi d/\lambda$) angle of scan. This is the visible range of the array antenna. The significant properties of $|AF(\psi)|$ are as follows:

1. The main beam direction ϕ_0 is obtained from the relation $AF(\psi) = 1.0$, which is what we get when ψ tends towards zero and the angles in the numerator and denominator of eqn (29) become very small. Now $\psi = 0$ when $kd \cos \phi_0 + \delta = 0$ or $\cos \phi_0 = -\delta/kd$.

(a) Broadside array $\phi_0 = \pm \pi/2$, i.e. $\delta = 0$, when in phase excitation.

(b) Endfire array $\phi_0 = 0$ for maximum radiation, i.e. $\delta = -kd \cos \phi_0 = -kd = -2\pi d/\lambda$.

Note that if we do a linear sketch of the array factor against ψ, we will get the peak always appearing at $\psi = 0$. However, if we do a linear sketch of the array factor against

ϕ, the peak will appear at the ϕ angle towards which the maximum of the array factor is pointed. As we change the electronic angle δ, the peak of the array factor will change in the ϕ direction, while at the same time the array factor will be symmetrical about the x-axis.

2. *Null location.* For nulls, we have $|AF(\psi)| = 0$, i.e. when

$$\frac{N\psi}{2} = \pm n\pi, \qquad n = 1, 2, 3, \dots . \tag{30}$$

3. *Width of main beam.* The first-null beam width (FNBW) is the angular width between first nulls for large N. The first nulls occur when

$$\frac{N\psi_{01}}{2} = \pm \pi \text{ or } \psi_{01} = \pm 2\pi / N . \tag{31}$$

(a) Broadside array ($\delta = 0$, $\phi_0 = \pi/2$):

$$\psi = kd \cos \phi .$$

First null at ϕ_{01} = width of main beam. $2\Delta\phi = 2(\phi_{01}-\phi_0)$. At ϕ_{01}, .

$$\cos \phi_{01} = \cos(\phi_0 + \Delta\phi) = \psi_{01} / kd$$

For $\phi_0 = \pi/2$,

$$\cos\left(\frac{\pi}{2} + \Delta\phi\right) = -\sin \Delta\phi = -2\pi / Nkd ,$$

$$\Delta\phi = \sin^{-1}(\lambda / Nd) \approx \frac{\lambda}{Nd} \qquad \text{for } Nd \gg \lambda . \tag{32}$$

With the broadside beam, there will be two main lobes, one with a maximum in the $\phi = \pi/2$ direction and the other with a maximum in the $\phi = -\pi/2$ direction. If we do not want one of the maximum beams (say in the $\phi = -\pi/2$ direction) by placing a reflecting metal plane of wire grid behind the antenna array (i.e. just below the x-axis) we could fold the beam in the $\phi = -\pi/2$ direction to point also in the $\phi = \pi/2$ direction. Thus the strength of the beam in the $\phi = \pi/2$ direction will be almost doubled.

(b) Endfire array ($\delta = -kd$, $\phi_0 = 0$):

$$\psi = kd(\cos \phi - 1) \quad \text{or} \quad \cos \phi_{01} - 1 = \frac{\psi_{01}}{kd} = -\frac{2\pi}{Nkd} = -\frac{\lambda}{Nd},$$

$$\cos \phi_{01} = \cos \Delta\phi \approx 1 - (\Delta\phi)^2 / 2 \quad \text{for small } \Delta\phi,$$

$$(\Delta\phi)^2 / 2 = \lambda / Nd \text{ or } \Delta\phi = \sqrt{\frac{2\lambda}{Nd}} , \tag{33}$$

since $Nd > \lambda/2$, i.e the FNBW of the endfire main beam is greater than the FNBW of the broadside main beam. The endfire beam could be designed to have one main beam in one direction only (e.g. in the $\phi = 0$ direction) without the use of a reflector.

In wireless mobile communications, base station antennas (normally mounted on top of buildings) use 12-element arrays to get a beam width of about 30° and 24 elements to get 15° beam width with a 120° reflector placed behind the array. The directivity D, or the gain G, of the array antennas can now be obtained from eqns (10) and (12) of Chapter 1, noting that the half-power beam width θ_B is approximately equal to FNBW/2, and hence $\theta_B = \Delta\phi$.

4. *Sidelobe locations.*   Sidelobes are minor maxima, side is a maximum. That is when

$$\left|\sin\left(\frac{N\psi}{2}\right)\right| = 1 \text{ or } N\psi/2 = \pm(2m+1)\frac{\pi}{2}, \quad m=1,2,\ldots \tag{34}$$

The first sidelobes occur when

$$\frac{N\psi}{2} = \pm\frac{3}{2}\pi \quad (m=1). \tag{35}$$

5. First sidelobe level

$$\frac{1}{N}\left|\frac{1}{\sin(3\pi/2N)}\right| \approx \frac{1}{N}\left|\frac{1}{3\pi/2N}\right| = \frac{2}{3\pi} = 0.212 \tag{36}$$

or $20 \log_{10}(1/0.212) = 13.5$ dB. These may be reduced by making the excitation amplitudes (i.e. peak currents) in the center elements higher than those in the end elements. The directivity for the N-element broadside is $D_b = 2N(d/\lambda) = 2(L/\lambda)$, where L is the length of the array antenna. The directivity of an N-element endfire array is $D_e = 4N(d/\lambda) = 4(L/\lambda)$. The directivity of an endfire array antenna is, in general, greater than that of a broadside array antenna. For a given directivity (or gain) we can determine the number of elements required in a linear array. The distance d between two adjacent elements is often limited by the presence of *grating* (or *ambiguity*) lobes periodically located at a distance inversely proportional to the distance d. They give rise to directional ambiguity, and could be eliminated by keeping d small. If the interelement distance d is reduced to too small a value, then problems related to mutual coupling, increased number of elements and increased cost of the antenna arise. Thus, keeping $d = \lambda/2$ has often been found to be a good compromise.

In general, if the antenna must scan between angles $-\theta_s$ and $+\theta_s$, then the interelement distance d must be kept to the following limit to prevent grating lobes appearing in the scanning region:

$$d < \lambda/(1 + \sin\theta_s), \tag{37}$$

where λ is the wavelength of the signal.

Figures 3.6–3.9 show the elemental pattern, array pattern and total radiation pattern on a five-element array antenna. Figure 3.7 is an endfire array, and Figures 3.6 and 3.8 show the pattern of a broadside array. Note that in either case, to get rid of one of the two main lobes, a reflector is placed behind the antenna to turn one of the main lobes over to add together with the other main lobe. An alternative way of reducing one of the two main lobes is to reduce the spacing below a half-wavelength, for instance to $d-$ 0.45λ. The idea here is that the visible region in the ψ domain is $2kd$, and to eliminate the grating lobe (the unwanted main lobe), the visible region should be reduced below the $d = 0.5\lambda$ value of 2π. Since $2\pi/N$ is the grating lobe half-width (maximum to null), we reduce the visible region by ensuring $2kd < 2\pi-(\pi/N)$, or $d < 0.5\lambda(1-0.5/N)$.

Example 2. Find the AF and plot the normalized radiation pattern of five isotropic elements placed $\lambda/2$ apart having excitation amplitude ratios $1 : 2 : 3 : 2 : 1$. Compare the first sidelobe level with a uniform five-element array.

$$
\begin{aligned}
\left|A(\psi)\right| &= \tfrac{1}{9}\left|1+2e^{j\psi}+3e^{j2\psi}+2e^{j3\psi}+e^{j4\psi}\right| \\
&= \tfrac{1}{9}\left|e^{j2\psi}\left(3+2\left(e^{j\psi}+e^{-j\psi}\right)\left(e^{j2\psi}+e^{-j2\psi}\right)\right)\right| \\
&= \tfrac{1}{9}\left|3+4\cos\psi+2\cos 2\psi\right|.
\end{aligned}
\tag{38}
$$

Broadside radiation: $\delta = 0$, $\psi = kd\cos\phi$, element spacing $d = \lambda/2$, $\psi = \pi\cos\phi$.

$$
\left|A(\phi)\right| = \tfrac{1}{9}\left|3+4\cos(\pi\cos\phi)+2\cos(2\pi\cos\phi)\right|.
\tag{39}
$$

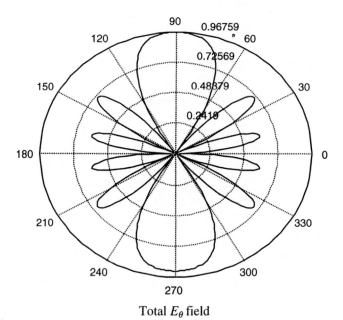

Total E_θ field

Figure 3.6: Total field (E_θ) pattern of a five-element array spaced at $\lambda/2$ apart ($\delta = 0$). Broadside array.

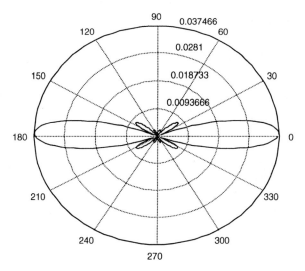

Figure 3.7: Element, array factor and total field (E_R) pattern of a five-element array spaced at $\lambda/2$ apart ($\delta = 0$). E_R is endfire.

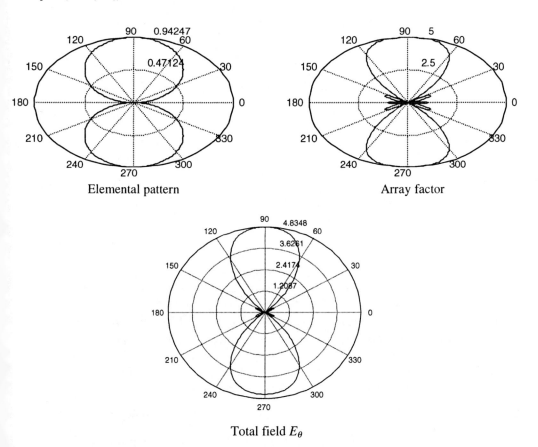

Elemental pattern

Array factor

Total field E_θ

Figure 3.8: Element, array factor and total field (E_θ) pattern of a five-element array spaced at $\lambda/2$ apart ($\delta = \pi$). Broadside array.

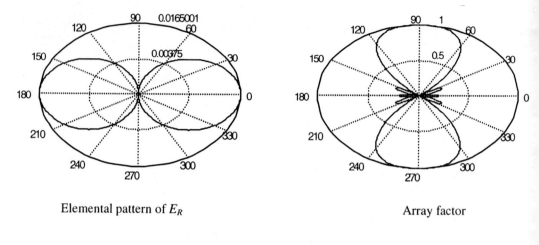

Elemental pattern of E_R Array factor

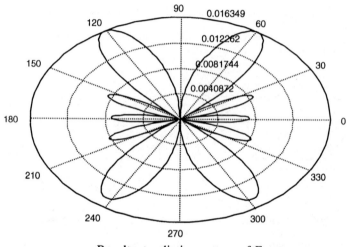

Resultant radiation pattern of E_R

Figure 3.9: Element, array factor and total field (E_R) pattern of a five-element array spaced at $\lambda/2$ apart ($\delta = \pi$).

The first sidelobe is at 0.11 or 19.2 dB down from main beam radiation. This compares with 0.25 or 12 dB down for the five-element uniform broadside array.

The above method of obtaining the resultant pattern of a collection of antennas is sometimes called pattern multiplication. For the two-element array antenna for instance, the resultant pattern is obtained by multiplying the radiation pattern of a single element (called the unit pattern), say that of a half-wave dipole, with the array factor, which is called the group pattern. Thus we have

$$\text{radiation pattern (RP)} = \text{unit pattern (UP)} \times \text{group pattern (GP)}. \qquad (40)$$

In the case of the two-element array made of half-wave dipoles, UP = radiation pattern of the half-wave dipole and GP = array factor. When we considered the N-element array, we had a simple analytical expression for the array factor. This method may be extended to more complex, non-linear, non-uniform arrangements too. What we must do is to break up a non-linear, non-uniform array into two sets of linear, uniform arrays. The first set would form just a two-element uniform, linear array called the unit. The distance between the two elements of the unit is d. The second set is called a group consisting of an $N/2$-element uniform, linear array. The distance between two adjacent elements in the group is $2d$. The elemental pattern of the unit and group arrays is the elemental pattern of the original array antenna, that is the elemental pattern of the actual antenna elements used (e.g. the elemental pattern of a half-wave dipole). Once we find the radiation pattern of the unit (UP) and that of the group (GP) separately, by multiplying the two we get the radiation pattern of the original array antenna.

For instance, consider six antenna elements arranged linearly in pairs of two elements. That is, there are three pairs of two-element arrays. Each two-element array may be made up of half-wave dipoles separated by a distance $\lambda/4$. Each pair of the two-element array may be separated by a center-to-center distance of $\lambda/2$. Now UP = radiation pattern of a two-element array, with the elements separated by $\lambda/4$, and GP = radiation pattern of three elements separated by a distance of $\lambda/2$. To obtain the radiation pattern of this three-element array, we use the array factor expression obtained for the linear N-element array, with $N = 3$. Once we find UP and GP, we may work out the radiation pattern RP. Such is the pattern multiplication method. As a second example consider a four-element array, with all elements carrying equal current and the interelement distance d. The phase angle of the currents in the first and third elements is 0, whereas that of the second and fourth elements is $\pi/2$. Thus the unit will consist of two elements separated by distance d, with the phases of the first and second elements being 0 and $\pi/2$, respectively. The group will also consist of two elements, but the interelement distance is $2d$ and the phase of both elements is 0. This method is little used today with the advent of the digital computer, which could very quickly work out the resultant field of complex antenna arrangements, given the radiation field of each element and the geometry of the array.

3.4   Mutual coupling between elements of the array antenna

We have thus far ignored the mutual coupling between elements placed in an array. The signal radiated by element number n, say, will hit the adjacent elements and be reflected back. If we consider dipole antenna elements, these will radiate in all directions and part of these radiated signals will hit and be reflected from the other elements. This means that these reflected signals would also be present due to mutual coupling between elements. So far we have only considered the direct waves uncontaminated by these coupled or reflected waves. The reflected waves could destructively or constructively interact with the direct waves. This gives rise to the presence of blind angles at which the array factor goes to zero due to reflected waves. We shall here obtain an expression for the reflection coefficient of an array antenna; this reflection coefficient will describe the ratio between the reflected waves and the direct waves incident on the elements of the array antenna.

Let the direct waves be p_i, $i = 1, 2, ..., N$, from an N-element array antenna. For a transmitting antenna these direct waves emerge from each of the N-element arrays. In a receiving antenna, these direct waves are those that are incident on each element of the array. Let the reflected waves due to the p_i waves be q_i, $i = 1, 2, ..., N$. If the coupling coefficients are C_{im} characterizing the matrix of coupling coefficients, we have

$$q_i = \sum_{m=1}^{N} C_{im}\, p_m. \tag{41}$$

For an array with an interelement distance d, and if the signal at the 0th element is p_0, the signal at the mth element is given by

$$p_m = p_0 \exp(-jkmd \sin \theta), \tag{42}$$

where the difference in signal path length between the 0th and mth element is $md \sin \theta$. Therefore the reflected signals are given by

$$q_i = p_0 \sum_{m=1}^{N} C_{im} \exp(-jkmd \sin \theta). \tag{43}$$

Therefore the direction of arrival dependent reflection coefficient, $\Gamma_i (\theta) = q_i/p_i$, is given by

$$\Gamma_i(\theta) = \sum_{m=1}^{N} C_{im} \exp(-jk(i-m)d \sin \theta). \tag{44}$$

The coupling coefficient C_{im}, which is a function of the interelement distance d_{im}, where $d_{im} = \text{Modulus}(i-m)d$, may be written as

$$C_{im} = c_0\, d_{im} \exp(-jkd_{im}), \tag{45}$$

where c_0 is a decreasing function of d_{im}, e.g. $c_0 = \rho \exp(-\kappa(i-m)d)$, where ρ and κ are constants.

Thus the active reflection coefficient may be rewritten as

$$\Gamma_i(\theta) = \sum_{m=1}^{N} c_0 d_{im} \exp[-jk(d_{im} + (i-m)\, d \sin \theta)]. \tag{46}$$

It is of interest to note that the active reflection coefficient approaches unity (i.e. total reflection) for certain values of θ. These values of θ are called blind angles since the receiver antenna will not capture any signal at these angles. The blind angles are close to the grating lobe angles, and are given by the relation

$$k(d_{im} + (i-m)d \sin \theta) = 2n\pi \quad (n = 1,2,...),$$

from which we get, with $n = 1$,

$$\sin \theta = \lambda/d - 1. \tag{47}$$

3.5   Polarization

In our presentation of antenna theory so far we have considered only the wire type of antennas. These antennas radiate linearly polarized signals. Thus for example, if we consider the half-wave dipole antenna radiation in the far field, the electric field is in the $\mathbf{u}_\theta$ direction, but if we consider the significant electric field strength (in the 3 dB range) then, within the 3 dB cone, unit vector is roughly parallel to u_z and the radiation field will be parallel to the line element. This is the reason why in mobile communication and in television transmission, with the transmission antenna kept vertical, the receiver antenna should also be kept vertical to capture the peak radiated signal strength. In amateur radio, where over the horizon communication links are formed by bouncing off 6–20 MHz signals off the ionosphere (charged particle layer at heights of about 70 km above the earth), horizontally placed antennas are used to transmit and receive horizontally polarized electric fields. However, in satellite communications frequencies are reused by sending out two signals (each carrying different voice and video signals) polarized perpendicular to each other. Thus the two orthonormal signals do not interfere with each other, although both are at the same frequency and travel over the same path simultaneously. Two signals, polarized perpendicularly to each other (say in the $\mathbf{u}_x$ and $\mathbf{u}_y$ directions), will produce a resultant signal, which will rotate in space as it travels over free space. This is called a circularly (or elliptically) polarized signal. To transmit and receive such orthonormal signals, we need antennas which can effectively handle signals that are perpendicular to each other and produce a resultant circularly polarized signal (see Fig. 3.10). Consider the resultant signal due to two orthonormal signals.

$$\begin{aligned}
\mathbf{E}_\theta &= \mathbf{u}_x E_1 \cos(\omega t - kr) + \mathbf{u}_y E_2 \cos(\omega t - kr - \pi/2) \\
&= \mathbf{u}_x E_1 \cos(\omega t - kr) + \mathbf{u}_y E_2 \sin(\omega t - kr) \\
&= \mathbf{u}_x E_x + \mathbf{u}_y E_y.
\end{aligned} \tag{48}$$

The resultant signal in this case rotates in the counterclockwise direction and is called a right-hand-polarized wave (RPW), since its motion is similar to rotating one's outstretched right hand into the body, over the head. At $r = 0$, we have

$$\cos \omega t = \frac{E_x}{E_1},$$

$$\sin \omega t = \frac{E_y}{E_2} = \sqrt{1 - \left(\frac{E_x}{E_y}\right)^2}.$$

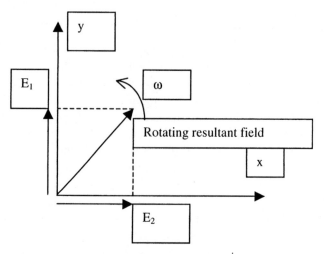

Figure 3.10: Circularly polarized signal.

Hence the trajectory of the resultant electric field is given by

$$\left(\frac{E_y}{E_2}\right)^2 + \left(\frac{E_x}{E_1}\right)^2 = 1.$$

If $E_1 = E_2$, the resultant signal rotates in a circle at a velocity equal to the electrical frequency as it propagates through space at the velocity of light. The direction of rotation is in the anticlockwise direction (right hand circularly polarized signal) or clockwise direction (left hand circularly polarized signal), depending on which signal is leading in phase. If the magnitudes are not equal then we get the resultant signal rotating over an ellipse, and it is called an elliptically polarized signal.

To transmit and receive such signals we use aperture antennas, which can also produce very narrow beams.

3.6   Aperture antennas

Aperture antennas are antennas in which the electromagnetic signals are radiated not from conduction currents flowing along wires, as in the case of wire antennas, but from displacement currents, which appear at the opening of a waveguide. Waveguides are hollow conductors, which transmit signals not by conduction currents, as in coaxial cables, but through the displacement currents existing in the empty space inside a hollow tube, normally shaped, in a rectangular hollow conductor. The waveguide is left open and the signals which travel along inside are launched out into space, thus the opening forming an aperture antenna. The shape of the opening differs; one popular shape is that of a horn, so that the waveguide is opened out to provide a wider aperture to radiate from.

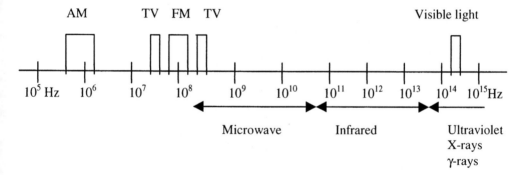

Figure 3.11: Microwave spectrum in which aperture antennas are largely used.

The aperture antennas work in the microwave frequency spectrum (see Fig. 3.11), roughly stretching from 0.1 GHz to about 300 GHz. Above this frequency of 0.3 trillion Hz, the electromagnetic spectrum is termed infrared, mostly associated with heat emission. Close to the visible light spectrum of 430×10^{12} to 750×10^{12} Hz, we have optical waveguides (largely made of glass) which transmit optical electromagnetic signals. Optical transmission is a major competitor to microwave transmission, with more and more undersea optical fiber cables interconnecting continents for telecommunication purposes. Above the visible light spectrum we have the ultraviolet electromagnetic signals, which carry enormous amounts of energy (given by Planck's constant $h \times$ frequency f) and are highly life threatening. A limited amount of wireless transmission may be achieved with laser beams traveling over free space; but their use will be highly localized, unlike a microwave satellite link, which can cover about one-third of the earth with one satellite in geostationary orbit. For wireless communications, the microwave region is the highest we can sensibly achieve, and as frequencies increase, newer types of antennas are required to handle these signals. Almost all antennas used at microwave frequencies are of the aperture type.

A rectangular aperture is shown in Fig. 3.12; this may simply represent a waveguide that is left open at one end or the aperture of a rectangular horn antenna. The signal comes into the aperture through the waveguide, and forms displacement currents ($\varepsilon \, dE/dt$ A/m^2, where ε is the permittivity of air) at the aperture. These displacement currents, like conduction currents along a linear array of dipole antennas, radiate out the signals. We would like to determine the electric field at observation point $P(x, y, z)$. Point P is assumed to be a point in the far-field region. The electric field $E(x, y, z)$ at point P is due to the displacement currents at the aperture. The displacement currents are associated with the electric field $E(x', y', 0)$ at the aperture, where the aperture is placed on the x–y-plane at $z = 0$.

Consider for a moment the electric field lines that appear at the aperture of the antenna. The electric field lines can be broken into small lines of fictitious elemental dipoles carrying current $\varepsilon \, dE/dt$ ($= j\varepsilon\omega E$) A/m^2. For each one of these little dipoles, we may determine the electric field radiated as (from Chapter 2)

$$E_\theta = j\eta \; \frac{kILe^{-jkr} \sin \theta}{4\pi r}, \tag{49}$$

which, for the fictitious dipoles, may be rewritten as

$$E_\theta = (jk/2\pi) \; \frac{\eta ILe^{-jkr} \sin \theta}{2r}. \tag{50}$$

In the case of the aperture antenna we replace $\eta I(L/2)$ by the aperture electric field $E_a = E(x',y',0)$ and integrate the electric field over the aperture to get the resultant radiated electric field. The magnetic field $H = E/\eta$.

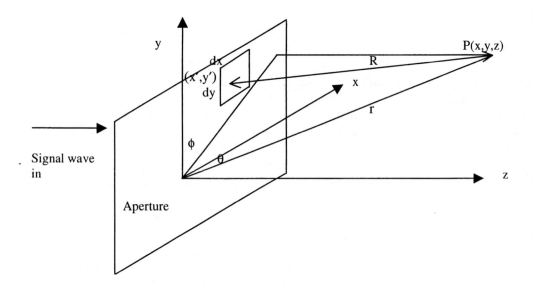

Figure 3.12: A radiating aperture.

A general expression relating the radiation field $E(x, y, z)$ to the field at the aperture is

$$E = \frac{jk}{2\pi} \int \frac{E(x',y',0)e^{-jkR}}{R} \, ds, \tag{51}$$

where $E(x', y', 0)$ is the scalar field in the aperture. The surface integration is carried over the aperture. This is a basic result of aperture theory for two-dimensional problems. It says that the radiation field pattern E is the Fourier transform of the aperture distribution $E(x', y', 0)$. The result is associated with Huygens' principle. According to this principle, the electric field $E(x', y', 0)$ and magnetic field intensity $H(x', y', 0)$ which appear at the aperture may be replaced by fictitious current sources. One model for the fictitious current sources is given by an electric conductor carrying current density of $-\mathbf{u}_n \times E(x', y', 0)$ replacing the aperture, where $\mathbf{u}_n$ is the unit vector normal to the aperture. In effect, we may picture the aperture as made up of a series current carrying conductors associated with lines perpendicular to the electric field lines on the same plane as the aperture.

For x', $y' \ll r$ the distance R between points at $(x', y', 0)$ and (x, y, z) is given by

$$
\begin{aligned}
R &= \sqrt{(x-x')^2 + (y-y')^2 + z^2} \\
&= \sqrt{r^2 - 2xx' - 2yy' + (x')^2(y')^2} \\
&= r\sqrt{\left[1 - \frac{2xx'}{r^2} - \frac{2yy'}{r^2} + \left(\frac{x'}{r}\right)^2 + \left(\frac{y'}{r}\right)^2\right]} \\
&\approx r\left(1 - \frac{xx'}{r^2} - \frac{yy'}{r^2}\right) \\
&= r - x'\sin\theta\cos\phi - y'\sin\theta\sin\phi,
\end{aligned}
\tag{52}
$$

where we have used $x = r\sin\theta\cos\phi$, $y = r\sin\theta\sin\phi$ and $(x'/r)^2$, $(y'/r)^2 \ll (xx'/r^2)$, (yy'/r^2).

It is this expression which will be substituted for the R appearing in the phase term inside the integral sign. Although $r \gg x'$, y' we cannot drop the last two terms, since even a small difference in distance will make a large difference in the signal phase for a high frequency signal. When we integrate over the aperture, the phase differences will either constructively or destructively sum together. For the R, which appears in the denominator, we may assume that $R \approx r$, since we are considering far away observation points, and the R in the denominator impacts the magnitude only.

$$
E(x, y, z) = \frac{jk}{2\pi r} e^{-jkr} \int E(x', y', 0) \; e^{jk\sin\theta(x'\cos\phi + y'\sin\phi)} dx'\,dy'.
\tag{53}
$$

We shall now evaluate this integral for an $a \times b$ rectangular aperture with a uniform field $E_0\mathbf{u}_y$ appearing in the aperture. This is a slightly idealized case since, in general, the electric fields inside a waveguide have a $\sin(\pi x/a)$ or $\sin(\pi y/b)$ term attached to E_0 (see Fig. 3.13).

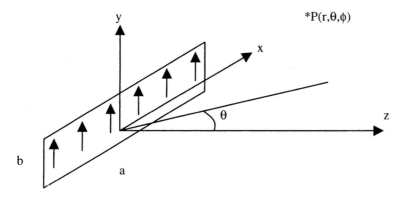

Figure 3.13: A rectangular aperture with uniform electric field.

We have $E(x', y', 0) = E_0 \, \mathbf{u}_y$ at $(x', y', 0)$ in the aperture. Hence the radiation field at point $P(r,\theta,\phi)$ is obtained using eqn (53):

$$
\begin{aligned}
E &= \frac{jke^{-jkr}}{2\pi r} \int\limits_{x'=-a/2}^{a/2} \int\limits_{y'=-b/2}^{b/2} E_0 e^{jk\sin\theta\,(x'\cos\phi + y'\sin\phi)} dx'\, dy' \\
&= \frac{jkE_0 e^{-jkr}}{2\pi r} \int\limits_{-a/2}^{a/2} e^{jx'k\sin\theta\cos\phi}\, dx' \int\limits_{-b/2}^{b/2} e^{jy'k\sin\theta\sin\phi}\, dy' \\
&= \frac{jkE_0 ab\, e^{-jkr}}{2\pi r} \left(\frac{\sin\psi_1}{\psi_1} \right) \left(\frac{\sin\psi_2}{\psi_2} \right),
\end{aligned}
\tag{54}
$$

where

$$
\psi_1 = \frac{ka\sin\theta\cos\phi}{2}
\tag{55}
$$

and

$$
\psi_2 = \frac{kb\sin\theta\sin\phi}{2}.
\tag{56}
$$

The terms $\sin\psi/\psi$ determine the radiation pattern of the aperture. Note that it is in some ways similar to the pattern function of a linear array. Thus the radiation pattern of an aperture is similar to the resultant pattern of two linear arrays arranged along the x and y axes. Indeed this is what we would expect if we place the lines of electric field lines of force at the aperture along which displacement currents flow. These displacement currents are line linear current elements arranged in the x and y directions at $z = 0$.

For $\phi = 0$, we get the E_θ radiation pattern in the zx-plane as

$$
E_{\phi=0} = \frac{kE_0 ab}{2\pi r} \left| \frac{\sin(ka\sin\theta)/2}{(ka\sin\theta)/2} \right|.
\tag{57}
$$

For $\phi = 90°$ we get the E_θ radiation pattern in the yz-plane as

$$
E_{\phi=90°} = \frac{kE_0 ab}{2\pi r} \left| \frac{\sin[(kb\sin\theta)/2]}{(kb\sin\theta)/2} \right|.
\tag{58}
$$

We may similarly obtain the E_ϕ pattern by setting $\theta = \pi/2$. The three-dimensional radiation pattern will contain both these patterns as well as each pattern for different values of ϕ.

$$
\textit{Note:} \quad \lim_{\psi \to 0} \frac{\sin\psi}{\psi} \to 1.
$$

We note the identical behavior in both planes (see Fig. 3.14) except for dimensions a and b, which will determine the width of the beams. Both E-plane and H-plane patterns will be unsymmetrical for the aperture antenna.

Beam width between first nulls may be obtained from

$$\frac{ka \sin \theta}{2} = \pi \tag{59}$$

or

$$\frac{ka\theta}{2} = \pi ,$$

assuming narrow beams. Hence

$$\theta = \frac{2\pi}{ka} = \frac{\lambda}{a} \text{ or } 2\theta = \frac{2\lambda}{a} = \text{FNBW}_{\phi=0} , \tag{60}$$

where FNBW is the first-null beam width. In general, for aperture antennas, the angular beam width is inversely proportional to the aperture size a (or diameter d for circular apertures) normalized to the wavelength λ. An approximate value for the half-power beam width (HPBW) is

$$\theta = \frac{\lambda}{a} = \text{HPBW}_{\phi=0} . \tag{61}$$

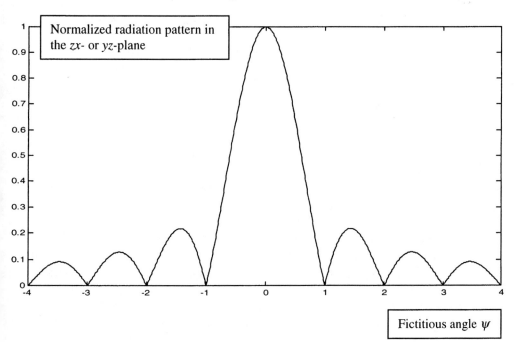

Figure 3.14: Two-dimensional aperture radiation pattern in the zx- or yz-plane.

The directivity of the antenna is given by

$$D = \frac{P_m}{P_r / 4\pi r^2} .$$ (62)

The total radiated power P_T is the power density at the aperture multiplied by the area of the aperture for uniformly distributed fields in the aperture. Hence the total radiated power is given by

$$P_r = \frac{1}{2} \frac{E_0^2}{\eta} (ab)$$ (63)

To find the maximum power density radiated P_m, we must first find the maximum field radiated:

$$E_{max} = j \frac{kE_0 ab \, e^{-jkr}}{2\pi r} .$$ (64)

Hence the maximum power density radiated by the rectangular aperture is given by

$$P_m = \frac{1}{2} \frac{|E_{max}|^2}{\eta} \text{ W/m}^2$$

$$= \frac{k^2 E_0^2 a^2 b^2}{8\pi^2 r^2 \eta} .$$ (65)

Hence the directivity of the aperture from eqns (62), (63) and (65) is given by

$$D = \frac{k^2 ab}{\pi} = \frac{4\pi}{\lambda^2} (ab) .$$ (66)

Note that the larger the aperture area (ab), the greater the directivity. This is one reason why large parabolic reflector dishes are used in microwave communications and in radio-astronomy, so as to increase both the directivity (D) and gain (G) of the antenna. A typical parabolic reflector will have a gain of about 60 dB, a diameter of 12 m for 4 GHz (satellite-to-earth) and 6 GHz (earth-to-satellite) space communications. Thus the physical size of the antenna is large compared to the millimeter or centimeter wavelengths involved. High gain or high directivity implies that the radiation pattern of the antenna can be synthesized with great accuracy. Space communications require very high gain antennas since distances are very large. The altitude of a geostationary satellite is 36,000 km, the distance between Earth and Moon is 360,000 km and the mean distance between Earth and Mars is 150 million km. As the frequency spectrum gets filled up, space communications at higher Ku-band frequencies (11–15 GHz) pose further challenges since atmospheric noise increases exponentially with frequency; this means that the antenna must be made to have low noise (cold antennas) and high gain.

uniform and given by $E_0 \cos(\pi y'/b)$, it could be shown that the radiation power will be halved, and directivity (or gain) $D = 32\ (ab)/(\pi \lambda^2)$ and effective aperture $A_{em} = 8(ab)/\pi^2$. Now the effective aperture is less than the physical aperture (ab).

One further point to note is that all along we have defined D and G for the maximum power density radiated. Therefore the gain in directions other than the direction of maximum power density radiation will be less than the G and D values we have obtained. If E is proportional to $\sin \theta$, for example, the gain and directivity will be functions of $(\sin \theta)^2$. The effective radiated power (ERP) of the antenna

$$\text{ERP} = P_r = G \cdot P_T, \tag{67}$$

where P_T is the power input to the antenna in W. If $G = 1000$ and is 5 W then ERP is 5000 W.

It should be remembered that electric field proper that appears at an aperture connected to a waveguide, denoted as $E(x', y', 0)$, may be obtained from waveguide solutions, e.g. for the TE_{10} mode.

$$E_y = -\frac{j\omega\mu\pi}{a\left(\gamma^2 + \omega^2/c^2\right)} H_{z0} \sin\left(\frac{\pi x}{a}\right). \tag{68}$$

A widely used aperture antenna is the horn antenna (often used together with the parabolic reflector) shown in Fig. 3.15. The directivity of a horn antenna, shown in Fig. 3.15, is given by

$$D = \frac{1}{2}\frac{4\pi(ab)}{\lambda^2}. \tag{69}$$

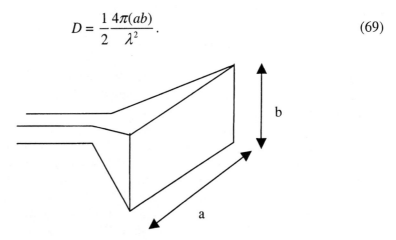

Figure 3.15: A rectangular horn antenna.

For a pyramidal horn the gain of the antenna is

$$G = 10A/\lambda^2 \approx 10(ab)/\lambda^2. \tag{70}$$

Its vertical −3 dB beam width is given by $\theta_v = 51\lambda/b$ and its horizontal −3 dB beam width $\theta_h = 70\lambda/a$. A 4×9 cm pyramidal horn is typically used at a frequency $f =$

Its vertical −3 dB beam width is given by $\theta_v = 51\lambda/b$ and its horizontal −3 dB beam width $\theta_h = 70\lambda/a$. A 4 × 9 cm pyramidal horn is typically used at a frequency $f = 10.25$ GHz. The vertical −3 dB beam width of a horn radiator is $\theta_v = 51\lambda/b$ and the horizontal −3 dB beam width is $\theta_h = 70\lambda/a$; both angles are given in degrees. Typical gain and beam widths that we obtained with horn antennas at 10 GHz are of the order of 50° and 25°, respectively. The parabolic reflector type of aperture antenna has a gain $G = 6d^2/\lambda^2$ and a −3 dB beam width (HPBW) of $70\lambda/d$ (degrees), where d is the diameter of the dish. The radiation pattern of such an antenna can be synthesized with a sampling step of λ/d. The primary feeds of such parabolic reflectors are horn antennas which direct the waves onto the surface of the parabolic reflectors. Thus we have two apertures here: that of the horn antenna and that of the reflector. The latter has a much larger aperture area leading to high gain. The typical efficiency of such antennas (i.e. $G/D = P_r/P_T$) is 60%. With structural improvements (e.g. Cassegrain antenna) it is possible to get gains of the order of 80%, but the cost of these antennas is very high. With parabolic dishes it is possible to get gains of the order of 10^5 and beam widths of the order of 0.1° at operating frequencies of 10 GHz and above. A microwave antenna with a gain of 10^5 and a noise temperature of 55 K (i.e. 17.4 dB with respect to 1 K) will have an antenna gain to total noise temperature ratio (called the *figure of merit*) of 32.6 dB. When a horn antenna is mounted outdoors, it is generally a good idea to cover the aperture with a plastic paper to prevent rainwater getting into the antenna. This is particularly critical with the low noise block converter (LNBC) electronics mounted just adjacent to the horn. The LNBC consists of a frequency downconverter and a low noise amplifier. Only a short strip of waveguide of a few centimeters long was needed to connect the horn antenna to the microwave input terminal of the LNBC. The LNBC is mounted outdoors next to the horn antenna in order to be able to downconvert the received microwave signal to a lower frequency before transmitting it over a coaxial cable into the radio room. In the older installations, the LNBC was placed indoors, which meant that the rather cumbersome waveguides had to be used to transmit the signal from the antenna into the radio room.

An aperture array antenna can be constructed by cutting a series of rectangular slots (apertures) on one of the four walls of a rectangular waveguide. In this case there will be radiation out of each rectangular slot, and the resultant waveform will be the summation of radiation from each hole or aperture. Such a 7.5 × 1.5 m array with 4000 slots with beam widths of 1° and 4.75° constitutes the AWACS (airborne warning and control system) antenna costing about US$ 300 million with the transmitter/receiver electronics; it is normally mounted on top of a 707 civilian aircraft.

3.7   Patch microstrip antenna

One type of aperture antenna, which has become very popular, is the microstrip printed or patch antenna. It is a small, light antenna, made of copper patches mounted on dielectric substrates, as shown in Fig. 3.16.

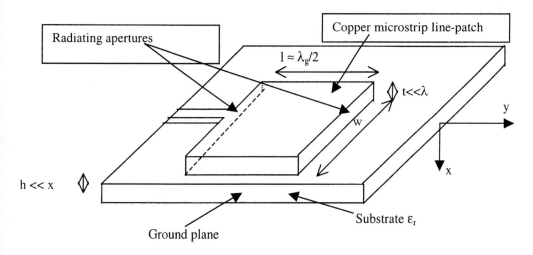

Figure 3.16: A rectangular microstrip patch antenna.

By using dielectric substrates of high permittivity, we make the wavelength short. This means that the antenna can be very small, since the size of the antenna depends on the frequency (or half-wavelength) of the signals coming into it. They are particularly attractive candidates for mobile, aircraft and spacecraft applications. These copper patches are flash mounted, with the feedline, which is also another microstrip line, placed behind the ground plane. These are inefficient radiators and have a narrow frequency bandwidth of about 5% of the center frequency f_0. The relatively narrow bandwidth is due to the patch antenna behaving as a resonator. There are also added disadvantages of low radiation efficiency and poor polarization purity, but the fact that these antennas may be realized with printed circuit technology and are suitable for electronic system miniaturization has made them attractive for telecommunications, radar, mobile communications, space industry and medical applications. Figure 3.16 shows a rectangular printed antenna, which is the simplest and most widely used geometry. Electromagnetic radiation takes place through the $w \times t$ aperture, with the beam pointing in the y direction. The radiating elements and the feedlines are usually photoetched on the dielectric substrate.

The antenna has two slots each $w \times h$, placed perpendicular to the feedline. Between the slots is a transmission line of length $l = \lambda_g/2$. Two slots array, each element separated by $\lambda_g/2$. Fields at each slot have opposite polarization ($\lambda_g/2$). The y components are out of phase and cancel out. Only the TEM mode exists in feedlines. The electric field inside the cavity-like patch antenna (see Fig. 3.17) is given by

$$E_z = E_0 \cos(m\pi x/a) \cos(n\pi y/b), \qquad (71)$$

where E_0 constitutes the magnitude and phase of the field and the constants m and n are related to the signal angular frequency and the velocity of the wave in the substrate u (where $u = (\varepsilon\mu)^{-1/2}$) by

$$(m\pi/a)^2 + (n\pi/b)^2 = (\omega/u)^2. \qquad (72)$$

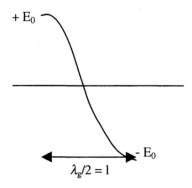

Figure 3.17: Electric field distribution at the aperture.

For $a > b$, the fundamental mode ($m = 1$. $n = 0$) fields inside the patch antenna (resonator) are given by

$$E_z = E_0 \ \cos(\pi x/a), \tag{73}$$
$$H_x = 0, \tag{74}$$
$$H_y = Z \ E_0 \sin(\pi x/a), \tag{75}$$

where

$$Z = (j/\omega\mu) \ (\pi/a). \tag{76}$$

In order to obtain the radiated field from these fields at the patch antenna aperture, we have to integrate these fields over the $a \times b$ aperture. To simplify our discussion, we shall assume uniform electric and magnetic fields over the aperture. Therefore the aperture electric field is given by

$$E_a = \mathbf{u}_x E_0 \begin{cases} -h/2 \ \le \ x' \ \le \ h/2, \\ -w/2 \ \le \ z' \ \le \ w/2. \end{cases} \tag{77}$$

Hence the magnetic current density at the aperture is given by

$$M_s = -2 \, \hat{n} \times \mathbf{E}_a = -2u_y \times u_x E_0 = u_z 2E_0, \tag{78}$$

and the electric current density is

$$J_s = 0. \tag{79}$$

The far-field region fields are given by

$$E_r = E_\theta = 0, \tag{80}$$
$$E_\phi = -j \frac{hwkE_0 \ e^{-jkr}}{2\pi r} \left\{ \sin \theta \left[\frac{\sin \psi_1}{\psi_1} \right] \left[\frac{\sin \psi_2}{\psi_2} \right] \right\}, \tag{81}$$

where

$$\psi_1 = \frac{kh}{2}\sin\theta\cos\phi, \tag{82}$$

$$\psi_2 = \frac{kw}{2}\cos\theta. \tag{83}$$

The simplified radiation electric field for the condition $h \ll \lambda$ is given by

$$E_\phi = -j\ \frac{V_0 e^{-jkr}}{\pi r}\sin\theta\frac{\sin\left[(kw/2)\cos\theta\right]}{\cos\theta}. \tag{84}$$

The voltage at the aperture is related to the field at the aperture by the following relation:

$$V_0 = hE_0. \tag{85}$$

Hence the gain of the antenna is given by

$$G = \frac{2P}{|V_0|^2} = \frac{I}{120\pi^2}, \tag{86}$$

where I is obtained from the radiation electric field as

$$I = \int_0^\pi \left[\frac{\sin\left((kw/2)\cos\theta\right)}{\cos\theta}\right]^2 \sin^3\theta\,d\theta. \tag{87}$$

Hence the patch antenna gain is given by

$$G = \begin{cases} \dfrac{1}{90}\left(\dfrac{w}{\lambda}\right)^2 & \text{for } w \ll \lambda, \tag{88} \\[3mm] \dfrac{1}{120}\left(\dfrac{w}{\lambda}\right)^2 & \text{for } w \gg \lambda. \tag{89} \end{cases}$$

The directivity of the antenna is

$$D = \left(\frac{2\pi w}{\lambda}\right)^2\frac{1}{I} \tag{90}$$

$$= \begin{cases} 3 & \text{for } w \ll \lambda\ (4.77\ \text{dB}), \\ 4 & \text{for } w \gg \lambda\ (6.02\ \text{dB}). \end{cases}$$

The bandwidth of the printed or patch antenna may be increased by:

(i)   increasing the impedance of the line or by increasing h, but increasing h will mean that the antenna is no more low profile;

(ii) using high ε_r substrate to reduce dimensions of parallel line;

(iii) increasing L of the microstrip by cutting holes or slots;

(iv) adding reactive components to reduce the VSWR at the antenna.

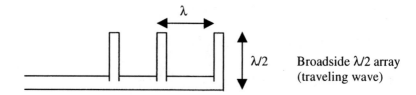

Broadside $\lambda/2$ array
(traveling wave)

Figure 3.18: Traveling wave microstrip antenna array.

Microstrip array antennas are shown in Figs 3.18 and 3.19. These are compact array antennas with high directivity. The compactness is achieved by reducing the wavelength using high permittivity substrate. These antennas may replace wire antennas in mobile communication systems.

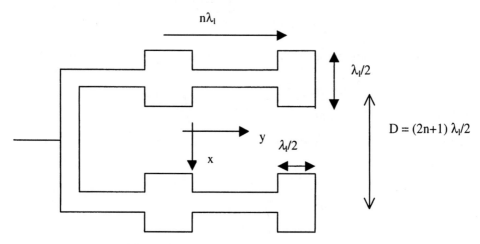

Figure 3.19: A popular patch antenna array.

The array spacing D suppresses radiation in the x direction, and helps to direct the antenna beam along the y direction. Such an antenna may be used in aircraft-to-satellite communications. Since the antenna is small and relatively flat it may be placed on top of the aircraft without causing wind friction. The two elements in the top row and the two elements in the bottom row can be electronically phase controlled to steer the beam in the azimuth direction. Such azimuth control will be required as the aircraft flies past the satellite with which it is maintaining communication. When it is flying towards the satellite, the beam will have to look forward. As it flies past the satellite position, the beam will have to be steered so that, after flying past, the beam looks back at the satellite antenna. Thus azimuth control of the microstrip antenna array is required to keep in contact with the satellite. The two rows of arrays, separated by D, will provide the elevation angle beam, since the aircraft cruise altitude (e.g. 10 km) will be much less than the altitude of the satellite (e.g. 36,000 km). The elevation angle beam need not be steered since the altitude difference between the aircraft and satellite will be constant when the aircraft is cruising.

3.8   Television receiver antennas

In this section we discuss the practical arrangements which are suitable for television receiver antennas. These are array antennas, and it will be seen that when the number of elements is increased both the gain and the directivity of the antenna will improve.

3.8.1   Code of practice for reception of television broadcasting

An antenna is a device comprising either a single element or an assembly of elements capable of converting the intercepted electromagnetic waves into an electromotive force at its terminals. An antenna element is a primary or secondary radiator that is a component of an antenna. A parasitic element is an antenna element that is not connected to a transmitter by a feeder. A lobe is a portion of the radiation pattern of an antenna that is contained within a region bounded by the direction of maximum radiation. The main lobe is the lobe containing the direction of maximum radiation. The position of the antenna is of considerable importance whether it is used as a transmitter or as a receiver. All local obstructions between the receiving antenna and the transmitter should be avoided as far as possible. Multiple images on the television screen may often be eliminated by a careful choice of position for the antenna and by a proper use of its directional properties. To study the effect of reflection, it is necessary to use equipment that will either give a picture display or provide the possibility of observing delayed secondary signals. It must be noted that the attention effect of trees on the television signal is significant and therefore should be avoided.

Where possible, the antenna should be mounted well clear of other conductors, including structural metal work and other antennae. Every effort should be made to install antennae as far as possible from power lines, to avoid danger from electrification.

In order to ensure satisfactory performance and durability of an outdoor antenna system, due regard should be paid to protection against any possible deterioration of components as a result of continued exposure to atmospheric variations. The materials selected for the antenna installation should be chosen so as to resist corrosion. Generally the antenna array is constructed of aluminum tubing and this has natural protection against corrosion. All other components, including masts and mounting brackets, should be made of steel. These parts should be protected by hot dip galvanizing or by zinc or cadmium plating. If dissimilar materials are brought into contact they should be selected or plated so as to minimize galvanic corrosion.

Most commonly, galvanized steel tubular masts are used to mount antenna arrays. In selecting the size of the mast, consideration should be given to the required height and loading due to the antenna and wind. The mast should have minimum wall thickness and maximum length for a given diameter. Where the antenna height is not excessive and the mast is of metal, it may be installed on the ground itself, which would eliminate the need for separate earthing.

When it is necessary to mount the mast at a height above the ground, it must be secured using a minimum of two brackets. Antenna masts of more than 6 m should be supported with a minimum of three guy wires positioned equiangularly and properly anchored to the ground or other suitable points. For every 6 m lengths of antenna mast there should be a set of guy wires, each set comprising a minimum of three.

Metal masts, supporting structures and the booms of television antenna should be connected to the earth to minimize the buildup of static charges on antenna installations, which cause damage, danger and interference. In most cases, it is sufficient to ensure that the mast is earthed, relying on the clamp that secures the antenna to the mast for electrical contact between the antenna and the mast. A copper conductor should be used for this earthing connection with a minimum cross sectional area of 1.5 mm^2.

3.8.2   VHF/UHF receiving antennas

A dipole antenna consists of two in-line rods or wires with a total length equal to a half-wave at the primary band it is intended to cover. Normally a 300 Ω balanced transmission line is connected at the center. However, since the radiation resistance of the half-wave dipole is 73 Ω, the use of the usual 300 Ω balanced cable will result in a mismatch and some loss in transfer of the received signal power. In addition, if the receiving end of the line is not a matching impedance of approximately 73 Ω, there will be reflections up and down the line. These will show up in the received picture as a horizontal smear for short lines or as ghosts displaced to the case of longer lines.

An increase in bandwidth and an increase in impedance may be accomplished also by use of the two-element folded-dipole configuration (see Figs 3.20 and 3.21).

The impedance can be increased by a factor of as much as 10 by using rods of different diameter and varying the spacing. The typical folded dipole with elements of equal diameter has a radiation impedance of 290 Ω, which is four times that of a standard dipole and closely matching the 300 Ω twin lead. The radiation pattern and directivity of a dipole is dependent on the length of the dipole. This is illustrated in Fig. 3.22 for sinusoidal signals fed to a horizontally placed dipole in free space. The most widely used dipole is the half-wave dipole; when the length is increased beyond this, the radiation pattern begins to break up as shown in Fig. 3.22. The length of the antenna L is related to its diameter D by $L/D = 37$ to 7 for 1 GHz operation, 150 (reception) or 30 (transmission) for 100 MHz operation, and 100 for 30 MHz operation.

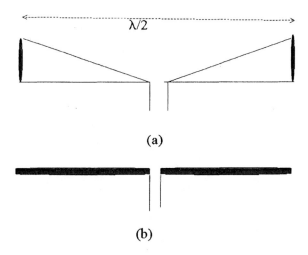

(a)

(b)

Figure 3.20: (a) Conical and (b) cylindrical half-wave dipole antennas.

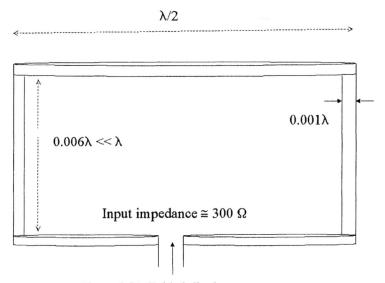

Figure 3.21: Folded-dipole antenna.

The Yagi–Uda dipole array antenna is a high gain, low cost, low wind resistance narrow-band antenna suitable for single TV channel reception. Such antennas are popular in remote locations where high gain is required or where only a few channels are to be received. An empirically verified design procedure for the design of Yagi–Uda dipole arrays which includes compensation of dipole element lengths for element and metallic-boom diameters is presented. Optimum Yagi–Uda arrays have one driven

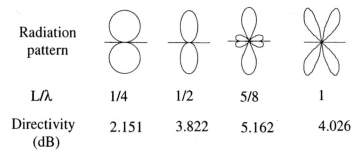

Radiation pattern				
L/λ	1/4	1/2	5/8	1
Directivity (dB)	2.151	3.822	5.162	4.026

Figure 3.22: Length dependent radiation patterns of dipole.

dipole, one reflecting parasitic dipole, and one or more directing parasitic dipoles as shown in Fig. 3.23. Since the input impedance of the Yagi–Uda antenna is much lower than that of an isolated dipole, it is the usual practice to use the folded dipole in order to push the input impedance closer to the 75 Ω impedance of the coaxial cable. A major limitation of the Yagi–Uda antenna is its narrow bandwidth, so that it is necessary to use different Yagi–Uda antennas for different TV bands or areas.

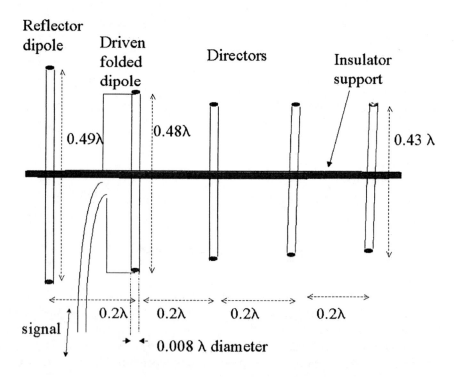

Figure 3.23: Yagi–Uda array antenna.

The design procedure is given for six different arrays. The driven dipole in all cases is a half-wavelength folded dipole which is empirically adjusted in length to achieve a minimum reflection at the design frequency. The length of the driven dipole has little impact on the gain of the array.

The beam width of the array depends on the number of elements in it. As the number of elements increases, the directivity improves. Typical characteristics are shown in Table 3.1.

Table 3.1: Typical characteristics of single-channel Yagi–Uda arrays.

No. of elements	Gain (dB)	Beam width (deg)
2	3–4	65
3	6–8	55
4	7–10	50
5	9–11	45
9	12–14	37
15	14–16	30

It is possible to design the Yagi–Uda antenna to operate as a broadband antenna. Arrays with no more than five or six elements can be broadbanded by shortening the directors for high frequency operation, lengthening the reflector for low frequency operation, and selecting the driven dipole for the mid-band operation.

The corner-reflector antenna is very useful for UHF reception because of its high gain, large bandwidth, low sidelobes, and high front-to-back ratio. A $90°$ corner reflector, constructed in grid fashion, is generally used (see Fig. 3.24).

For the corner reflector of Fig. 3.24, let d be the spacing between the half-wave dipole radiator and the corner. Let the half-wave dipole wire and the reflector wires all be parallel to the z-axis. It is possible to work out the resultant radiation field by considering an image of the dipole on the upper reflector plane, a second image on the lower reflector plane and a third image at the corner of the reflector. The contribution of the two-element array formed by the feed element and its image at the corner will be $2 \cos(kd \cos \phi)$ and the contribution from the two images in the reflector planes will be $2 \cos[kd \cos(\pi/2-\phi)]$. The resultant radiation field in the xy- or H-plane in the region $-\pi/4 \le \phi \le \pi/4$ is given by

$$AF(\theta = \pi/2, \phi) = 2[\cos(kd \cos \phi) - \cos(kd \sin \phi)], \qquad (91)$$

and the xz- or E-plane pattern is given by

$$AF(\theta, \phi = 0) = 2[-1 + \cos(kd \sin \theta)] \left| \frac{\cos((\pi/2) \cos \theta)}{\sin \theta} \right|. \qquad (92)$$

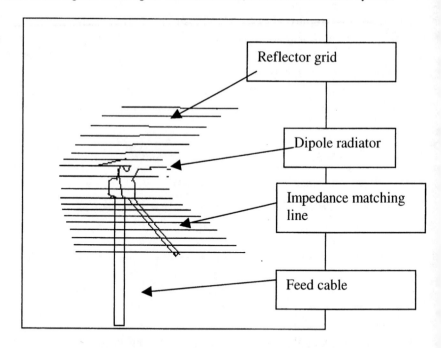

Figure 3.24: A corner reflector and a dipole antenna.

A wide-band triangular dipole of flare angle $40°$, which is bent $90°$ along its axis so that the dipole is parallel to both sides of the reflector, is used as the feed element. The dipole has a spacing of about (wavelength/2) at mid-band from the vertex of the corner reflector, i.e. $d = \lambda/2$. However, for this choice of d the input impedance of the antenna is quite high, close to 125 Ω. To bring down the input impedance without too great a drop in the gain, in most practical designs, $d = 0.35\lambda$ is used. Grid length L and grid width W should be kept about $L = 2d$ and $W = 1.5 \times$ length of the antenna feed. Hence if a half-wave dipole feed is used, $W = 1.5 \times 0.5\lambda = 0.75\lambda$. Keeping $W >$ length of the feed element ensures that there will be no radiation leaking into the back region. The spacing for grid tubing of 0.1λ diameter should be slightly under 0.5λ. An alternative to the dipole feed is the bow-tie antenna element, which has good impedance bandwidth properties.

3.9   Finite length antenna: a basic building block for antenna simulation

We have seen that some of the most complex antennas like array antennas and aperture antennas may be thought of as being made up of discrete line elements. In aperture antennas, for instance, the radiating line elements are the electric field lines that appear at the aperture carrying displacement currents. In electromagnetic image reconstruction too, the line element is expected to yield image reconstruction in shorter time and it inherently contains more information about the region it covers. This includes information about the size of the region and its angle of inclination. The equation of the electric field of a finite line element forms the basis of the imaging model. This model is developed and implemented in this report. This chapter will first set out to derive the equations for the electric field. Studies of the variation of the electric field with the various parameters in the field equation are performed to understand the impact of the model when used in image reconstruction.

Once the radial and tangential fields for a finite length radiator are derived, the structures of the radiation patterns are studied. These radiation patterns have a direct impact on the image quality, and also on the relative value of using measurements of the radial component, E_r, tangential components, E_θ, or magnitude of the electric field, $\sqrt{(E_r^2 + E_\theta^2)}$, for imaging in medicine and radar cross section measurements. This is an additional advantage that the line scatterer model provides over point scatterers, i.e. the use of three different signals for image reconstruction.

3.9.1 Derivation of electric field radiated by a finite line element

To derive the electric field for the line element, let us first consider an infinitesimal element of length h, carrying current as shown in Fig. 3.25.

Let $[I]$ be the retarded current carried by the element and $[Q]$ be the retarded electric charge. Both quantities are a function of $(t - R/c)$, where t is time, R is the distance from the point P to the center of line element and c is the speed of light. $[Q]$ is related to $[I]$ by the following equation:

$$\frac{d[Q]}{dt} = [I]. \qquad (93)$$

The three equations governing electromagnetic field are

$$\mathbf{E}(R,t) = -\nabla \mathbf{V}(R,t) - \frac{\partial \mathbf{A}(R,t)}{\partial t}, \qquad (94)$$

$$\mathbf{B}(R,t) = \nabla \times \mathbf{A}(R,t), \qquad (95)$$

$$\nabla \cdot \mathbf{A}(R,t) = -\frac{1}{c^2}\frac{\partial \mathbf{V}(R,t)}{\partial t}. \qquad (96)$$

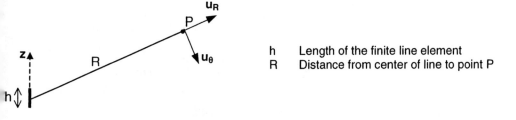

h	Length of the finite line element
R	Distance from center of line to point P

Figure 3.25: Orientation of an infinitesimal element carrying current.

The magnetic potential, **A**, is given as

$$\mathbf{A} = \frac{\mu_0}{4\pi} \int_v \frac{[\mathbf{J}]}{r} dv .\tag{97}$$

From the geometry of the problem in Fig. 3.25, eqn (97) can be expressed as

$$\mathbf{A} = \mathbf{u}_z \frac{\mu_0}{4\pi} \int_{-h/2}^{h/2} \frac{[I]}{R} dz.\tag{98}$$

Assuming that $R \gg h$, eqn (98) is approximated to

$$\mathbf{A} = \mathbf{u}_z \frac{\mu_0 h}{4\pi R}[I].\tag{99}$$

Expressing this in spherical coordinates

$$\mathbf{A} = \mathbf{u}_R \frac{\mu_0[I]}{4\pi R} h \cos\theta - \mathbf{u}_\theta \frac{\mu_0[I]}{4\pi R} h \sin\theta.\tag{100}$$

Divergence of **A** in spherical coordinates is given by

$$\nabla \cdot \mathbf{A} = \frac{1}{R} \frac{\partial}{\partial R} \left(R^2 A_R \right) + \frac{1}{R \sin\theta} \frac{\partial}{\partial}\left(A_\theta \sin\theta \right).\tag{101}$$

The first terms in the divergence **A** can be expressed as

$$\frac{\partial}{\partial R}\left(R^2 A_R \right) = \frac{\mu_0 h \cos\theta}{4\pi}\left(\frac{\partial[I]}{\partial R} R + [I] \right).$$

As $[I] = I(\theta)$, where $\theta = t - R/c$,

$$\frac{\partial[I]}{\partial R} = \frac{\partial[I]}{\partial\theta}\frac{d\theta}{dR} = \frac{\partial[I]}{\partial t}\frac{dt}{d\theta}\frac{d\theta}{dt}$$

$$= -\frac{1}{c}\frac{\partial[I]}{\partial t}.$$

Hence,

$$\frac{\partial}{\partial R}\left(R^2 A_R \right) = \frac{\mu_0 h \cos\theta}{4\pi}\left(-\frac{R}{c}\frac{\partial[I]}{\partial t} + [I] \right).$$

Following the same procedure the second term is found to be

$$\frac{\partial}{\partial\theta}(A_\theta \sin\theta) = -\frac{\partial}{\partial\theta}\left(\frac{\mu_0[I]}{4\pi R}h\sin^2\theta\right)$$

$$= -\frac{\mu_0[I]}{4\pi R}(2h\sin\theta\cos\theta).$$

Hence eqn (101) becomes

$$\nabla\cdot\mathbf{A} = \frac{\mu_0 h\cos\theta}{4\pi}\left(-\frac{1}{Rc}\frac{d[I]}{dt}-\frac{[I]}{R^2}\right). \tag{102}$$

From eqn (96),

$$\frac{\partial V(R,t)}{\partial t} = -c^2\nabla\cdot\mathbf{A}(R,t).$$

Integrating both sides with respect to time t,

$$V(R,t) = -c^2\int\nabla\cdot\mathbf{A}(R,t)\,dt.$$

Assuming that when $t<0$, the integral is zero, i.e. no potential before $t=0$,

$$V(R,t) = -c^2\int_0^t\frac{\mu_0 h\cos\theta}{4\pi}\left(-\frac{1}{Rc}\frac{d[I]}{dt}-\frac{[I]}{R^2}\right)dt$$

$$= \frac{c^2\mu_0 h\cos\theta}{4\pi}\left(\frac{[I]}{Rc}+\frac{1}{R^2}\int_0^t[I]\,dt\right).$$

Now, from the relationship of $[I]$ and $[Q]$ in eqn (93),

$$V(R,t) = \frac{c^2\mu_0 h\cos\theta}{4\pi}\left(\frac{[I]}{Rc}+\frac{[Q]}{R^2}\right). \tag{103}$$

After finding the $V(R,t)$ and $A(R,t)$, the electric field of the line element can be found by substituting these terms in eqn (94). Using

$$\nabla V(R,t) = u_R\frac{\partial V}{\partial R}+u_\theta\frac{1}{R}\frac{\partial V}{\partial\theta}+u_\phi\frac{1}{R\sin\theta}\frac{\partial V}{\partial\phi},$$

and letting

$$k = -\frac{c^2\mu_0 h}{4\pi},$$

we have

$$\frac{\partial V}{\partial R} = \frac{\partial}{\partial R} \left[k \cos \theta \left(\frac{[I]}{Rc} + \frac{[Q]}{R^2} \right) \right]$$

$$= k \cos \theta \left(\frac{d[I]}{dR} \frac{1}{Rc} - \frac{[I]}{R^2 c} + \frac{d[Q]}{dR} \frac{1}{R^2} - \frac{2[Q]}{R^3} \right)$$

$$= k \cos \theta \left(-\frac{1}{Rc^2} \frac{d[I]}{dt} - \frac{[I]}{R^2 c} + \frac{d[Q]}{dt} \frac{1}{R^2 c} - \frac{2[Q]}{R^3} \right)$$

$$= k \cos \theta \left(-\frac{1}{Rc^2} \frac{d[I]}{dt} - \frac{2[I]}{R^2 c} - \frac{2[Q]}{R^3} \right),$$

$$\frac{\partial V}{\partial \theta} = \frac{\partial}{\partial \theta} \left[k \cos \theta \left(\frac{[I]}{Rc} + \frac{[Q]}{R^2} \right) \right]$$

$$= -k \sin \theta \left(\frac{[I]}{Rc} + \frac{[Q]}{R^2} \right),$$

$$\frac{\partial V}{\partial \phi} = 0.$$

Hence,

$$\nabla V(R,t) = u_R \left[k \cos \theta \left(-\frac{1}{Rc^2} \frac{d[I]}{dt} - \frac{2[I]}{R^2 c} - \frac{2[Q]}{R^3} \right) \right] + u_\theta \left[-k \sin \theta \left(\frac{[I]}{Rc} + \frac{[Q]}{R^2} \right) \right],$$

$$\frac{\partial \mathbf{A}(R,t)}{\partial t} = u_R \frac{\mu_0 h \cos \theta}{4\pi R} \frac{d[I]}{dt} - u_\theta \frac{\mu_0 h \sin \theta}{4\pi R} \frac{d[I]}{dt}.$$

Since

$$\mathbf{E}(R,t) = -\nabla V(R,t) - \frac{\partial \mathbf{A}(R,t)}{\partial t},$$

we have

$$\mathbf{E}(R,t) = \mathbf{u}_R \left[k \cos \theta \left(\frac{1}{Rc^2} \frac{d[I]}{dt} + \frac{2[I]}{R^2 c} + \frac{2[Q]}{R^3} \right) \right] + \mathbf{u}_\theta \left[k \sin \theta \left(\frac{[I]}{Rc} + \frac{[Q]}{R^2} \right) \right]$$

$$- \mathbf{u}_R \frac{\mu_0 h \cos \theta}{4\pi R} \frac{d[I]}{dt} + \mathbf{u}_\theta \frac{\mu_0 h \sin \theta}{4\pi R} \frac{d[I]}{dt}$$

$$= \mathbf{u}_R \left[\frac{c^2 \mu_0 h \cos \theta}{4\pi} \left(\frac{1}{Rc^2} \frac{d[I]}{dt} + \frac{2[I]}{R^2 c} + \frac{2[Q]}{R^3} \right) - \frac{\mu_0 h \cos \theta}{4\pi R} \frac{d[I]}{dt} \right]$$

$$+ \mathbf{u}_\theta \left[\frac{c^2 \mu_0 h \sin \theta}{4\pi} \left(\frac{[I]}{Rc} + \frac{[Q]}{R^2} \right) + \frac{\mu_0 h \sin \theta}{4\pi R} \frac{d[I]}{dt} \right]$$

$$= \mathbf{u}_R \frac{h\cos\theta}{4\pi}\left[\left(\frac{\mu_0}{\varepsilon_0}\right)^{1/2}\frac{2[I]}{R^2} + \frac{2[Q]}{\varepsilon_0 R^3}\right]$$

$$+ \mathbf{u}_\theta \frac{h\sin\theta}{4\pi}\left[\left(\frac{\mu_0}{\varepsilon_0}\right)^{1/2}\frac{[I]}{R^2} + \frac{[Q]}{\varepsilon_0 R^3} + \frac{\mu_0}{R}\frac{d[I]}{dt}\right]. \qquad (104)$$

Transforming eqn (104) into cylindrical coordinates, taking into account

$$\cos\theta = \frac{z}{\left(r^2+z^2\right)^{1/2}}, \qquad \sin\theta = \frac{r}{\left(r^2+z^2\right)^{1/2}}, \qquad R^2 = r^2 + z^2,$$

$$\mathbf{E}(r,z,t) = \left(E_R\cos\theta - E_\theta\sin\theta\right)\mathbf{u}_z + \left(E_R\sin\theta + E_\theta\cos\theta\right)\mathbf{u}_r,$$

$$\mathbf{E}_z = \mathbf{u}_z \frac{h}{4\pi}\left[\left(\frac{\mu_0}{\varepsilon_0}\right)^{1/2}\frac{2z^2 - r^2}{\left(r^2+z^2\right)^2}[I] + \frac{1}{\varepsilon_0}\frac{2z^2-r^2}{\left(r^2+z^2\right)^{2.5}}[Q] - \mu_0\frac{r^2}{\left(r^2+z^2\right)^{1.5}}\frac{d[I]}{dt}\right],$$

$$\mathbf{E}_r = \mathbf{u}_r \frac{h}{4\pi}\left[\left(\frac{\mu_0}{\varepsilon_0}\right)^{1/2}\frac{3rz}{\left(r^2+z^2\right)^2}[I] + \frac{1}{\varepsilon_0}\frac{3rz}{\left(r^2+z^2\right)^{2.5}}[Q] + \mu_0\frac{rz}{\left(r^2+z^2\right)^{1.5}}\frac{d[I]}{dt}\right].$$

We shall assume that the finite line element is made up of infinitesimal line element with electric field dE and length dz (see Fig. 3.26).

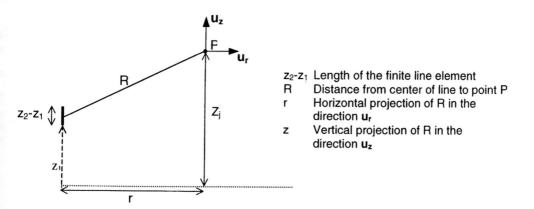

Figure 3.26: Orientation of a finite line element of length z_2-z_1.

Letting $h = dz$, $z = z_j - z$, and $dz = -dz$, the resultant electric field due to a line of length (z_2-z_1) carrying a current I is given by

$$\overline{\mathbf{E}} = \int_{z_1}^{z_2} \mathbf{E}(r, z, t) \, dz$$

$$= -\mathbf{u}_z \int_{z_1}^{z_2} \frac{1}{4\pi} \left[\begin{array}{c} \left(\dfrac{\mu_0}{\varepsilon_0}\right)^{1/2} \dfrac{2(z_j - z)^2 - r^2}{\left(r^2 + (z_j - z)^2\right)^2} [I] \\[4mm] + \dfrac{1}{\varepsilon_0} \dfrac{2(z_j - z)^2 - r^2}{\left(r^2 + (z_j - z)^2\right)^{2.5}} [Q] - \mu_0 \dfrac{r^2}{\left(r^2 + (z_j - z)^2\right)^{1.5}} \dfrac{d[I]}{dt} \end{array} \right] dz$$

$$- \mathbf{u}_r \int_{z_1}^{z_2} \frac{1}{4\pi} \left[\begin{array}{c} \left(\dfrac{\mu_0}{\varepsilon_0}\right)^{1/2} \dfrac{3r(z_j - z)}{\left(r^2 + (z_j - z)^2\right)^2} [I] \\[4mm] + \dfrac{1}{\varepsilon_0} \dfrac{3r(z_j - z)}{\left(r^2 + (z_j - z)^2\right)^{2.5}} [Q] + \mu_0 \dfrac{r(z_j - z)}{\left(r^2 + (z_j - z)^2\right)^{1.5}} \dfrac{d[I]}{dt} \end{array} \right] dz.$$

Using the integrals in Appendix 3.1, the integration of the terms for electric field can be evaluated. The final form of the electric field is given by the following equation. The r-component of the electric field is given as

$$\overline{\mathbf{E}}_r = -\mathbf{u}_r \left[\begin{array}{c} \dfrac{3r}{8\pi} \left(\dfrac{\mu_0}{\varepsilon_0}\right)^{1/2} \left\{ \dfrac{1}{\left(r^2 + (z_j - z_2)^2\right)} - \dfrac{1}{\left(r^2 + (z_j - z_1)^2\right)} \right\} [I] \\[5mm] + \dfrac{r}{4\pi\varepsilon_0} \left\{ \dfrac{1}{\left(r^2 + (z_j - z_2)^2\right)^{1.5}} - \dfrac{1}{\left(r^2 + (z_j - z_1)^2\right)^{1.5}} \right\} [Q] \\[5mm] + \dfrac{\mu_0 r}{4\pi} \left\{ \dfrac{1}{\sqrt{r^2 + (z_j - z_2)^2}^2} - \dfrac{1}{\sqrt{r^2 + (z_j - z_1)^2}^2} \right\} \dfrac{d[I]}{dt} \end{array} \right]. \quad (1\bullet$$

The z-component of the electric field is given as

$$\bar{E}_z = -\mathbf{u}_z \begin{bmatrix} \dfrac{1}{8\pi}\left(\dfrac{\mu_0}{\varepsilon_0}\right)^{1/2} \left\{ \dfrac{3\left(z_j - z_2\right)}{r^2 + \left(z_j - z_2\right)^2} - \dfrac{3\left(z_j - z_1\right)}{r^2 + \left(z_j - z_1\right)^2} \right. \\ \left. -\dfrac{1}{r}\left[\tan^{-1}\left(\dfrac{z_j - z_2}{r}\right) - \tan^{-1}\left(\dfrac{z_j - z_1}{r}\right)\right] \right\} \cdot [I] \\[2ex] + \dfrac{1}{4\pi\varepsilon_0}\left\{ \dfrac{z_j - z_2}{\left(r^2 + \left(z_j - z_2\right)^2\right)^{1.5}} - \dfrac{z_j - z_1}{\left(r^2 + \left(z_j - z_1\right)^2\right)^{1.5}} \right\}[Q] \\[2ex] + \dfrac{\mu_0}{4\pi}\left\{ \dfrac{z_j - z_2}{\sqrt{r^2 + \left(z_j - z_2\right)^2}} - \dfrac{z_j - z_1}{\sqrt{r^2 + \left(z_j - z_1\right)^2}} \right\}\dfrac{d[I]}{dt} \end{bmatrix}. \tag{106}$$

3.9.2   Electric field radiated by a line element

3.9.2.1   The generalized equations

Based on the line orientation geometry as shown in Fig. 3.27, $z_j = z$, $z_1 = 0$, and $z_2 = L$. The retarded charge $[Q]$ is assumed to be zero. Hence eqns (105) and (106) simplify to

$$E(r,z,t) = -\mathbf{u}_r \left\{ \dfrac{3r}{8\pi}\sqrt{\dfrac{\mu_0}{\varepsilon_0}} \left[\dfrac{1}{r^2 + (z-L)^2} - \dfrac{1}{r^2 + z^2} \right][I] \right.$$

$$\left. + \dfrac{\mu_0 r}{4\pi}\left[\dfrac{1}{\sqrt{r^2 + (z-L)^2}} - \dfrac{1}{\sqrt{r^2 + z^2}} \right]\dfrac{dI}{dt} \right\}$$

$$-\mathbf{u}_z \left\{ \dfrac{1}{8\pi}\sqrt{\dfrac{\mu_0}{\varepsilon_0}} \left[\dfrac{3(z-L)}{r^2 + (z-L)^2} - \dfrac{3z}{r^2 + z^2} - \dfrac{1}{r}\left(\tan^{-1}\left(\dfrac{z-L}{r}\right) - \tan^{-1}\left(\dfrac{z}{r}\right)\right) \right][I] + \right.$$

$$\left. \dfrac{\mu_0}{4\pi}\left[\dfrac{z-L}{\sqrt{r^2 + (z-L)^2}} + \dfrac{z}{\sqrt{r^2 + z^2}} \right]\dfrac{dI}{dt} \right\}.$$

$$\tag{107}$$

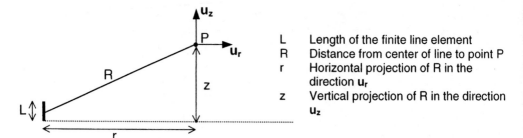

Figure 3.27: Orientation of a finite line element set up for investigation.

L	Length of the finite line element
R	Distance from center of line to point P
r	Horizontal projection of R in the direction $\mathbf{u}_r$
z	Vertical projection of R in the direction $\mathbf{u}_z$

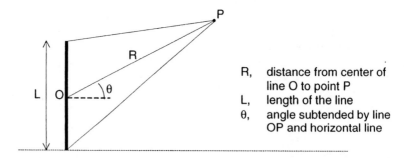

R,	distance from center of line O to point P
L,	length of the line
θ,	angle subtended by line OP and horizontal line

Figure 3.28: Finite line element in polar coordinates.

We can write eqn (107) in polar coordinates, a form which is used in the studies presented herein.

From Fig. 3.28, the equation of the electric field becomes

$$
\mathbf{E}(R,\theta,t) = -\mathbf{u}_r \left\{
\begin{aligned}
&\frac{3R\cos\theta}{8\pi}\sqrt{\frac{\mu_0}{\varepsilon_0}}\left[\frac{1}{R^2+(L/2)^2-RL\sin\theta}-\frac{1}{R^2+(L/2)^2+RL\sin\theta}\right][I] \\
&+\frac{\mu_0 R\cos\theta}{4\pi}\left[\frac{1}{\sqrt{R^2+(L/2)^2-RL\sin\theta}}-\frac{1}{\sqrt{R^2+(L/2)^2+RL\sin\theta}}\right]\frac{dI}{dt}
\end{aligned}
\right\}
$$

$$
-\mathbf{u}_z \left\{
\begin{aligned}
&\frac{1}{8\pi}\sqrt{\frac{\mu_0}{\varepsilon_0}}\left[\frac{3(R\sin\theta-L/2)}{R^2+(L/2)^2-RL\sin\theta}-\frac{3(R\sin\theta+L/2)}{R^2+(L/2)^2+RL\sin\theta}\right. \\
&\left.-\frac{1}{R\cos\theta}\left(\tan^{-1}\left(\frac{R\sin\theta-L/2}{R\cos\theta}\right)-\tan^{-1}\left(\frac{R\sin\theta+L/2}{R\cos\theta}\right)\right)\right][I] \\
&+\frac{\mu_0}{4\pi}\left[\frac{R\sin\theta-L/2}{\sqrt{R^2+(L/2)^2-RL\sin\theta}}+\frac{R\sin\theta+L/2}{\sqrt{R^2+(L/2)^2+RL\sin\theta}}\right]\frac{dI}{dt}
\end{aligned}
\right\}.
$$

$$(108)$$

3.9.2.2   The electric field components

From eqns (107) and (108), it is observed that the electric field of a finite length line element is composed of two components: the $\mathbf{u}_r$ component and the $\mathbf{u}_z$ component. The $\mathbf{u}_r$ component exists both in far-field and near-field regions, while the $\mathbf{u}_z$ component exists only in the near-field region. In this section, the following simulations are carried out to study the variation and structure of the electric field with distance:

(a) Variation of near-field $\mathbf{E}_z$ component with distance R from the center of the line element.
(b) Variation of far-field $\mathbf{E}_r$ component with distance R from the center of the line element.
(c) Variation of total electric field $\mathbf{E} = \sqrt{(\mathbf{E}_r^2 + \mathbf{E}_\theta^2)}$ with distance R from the center of the line element.

In all the simulation studies, the line element is assumed to have length $L = 1$ m. The retarded current $[I]$ is a sine function of the form $A \sin(2\pi ft)$, where A is the amplitude set equal to 1, and f is the signal frequency of value 1000 MHz. The magnitude of the electric field is dependent on the frequency. The $[dI]$ component in the equations of the electric field is proportional to the frequency. Therefore, as frequency increases the $[dI]$ component becomes dominant and may alter the intensity pattern of $\mathbf{E}_r$ and thus the total electric field $\mathbf{E}$. For the near-field component, $\mathbf{E}_z$, increasing the frequency increases the magnitude, since the effect of the $[I]$ component in $\mathbf{E}_z$ is always not dominant.

Variation of the E_z component with distance R.   Referring to Fig. 3.28, the distance R is varied from 0.4 to 1.5 m and electric field polar plots are plotted in Fig. 3.29. Each of the plots in Fig. 3.29 shows that the intensity of the near-field component at a fixed distance R varies with the angular position. The maximum intensity of the field occurs at the tip of the four lobes observed in each plot. This indicates that the near field of a finite line element shows some sort of directivity in its propagation. The intensity is 0 at angular positions of 0°, 90°, 180° and 270°.

At the distance $R = 0.5$ m, the pattern of the plot significantly changes. The intensity at angular positions 90° and 270° is the maximum. This is because the two angular positions at $R = 0.5$ m correspond to the two end points of the line element. A closer look at how the field intensity changes as the distance R approaches 0.5 m revealed that the intensity of the field is highest around the two end points. This is shown in Fig. 3.30.

It is also seen that as the distance R is increased the intensity of the field decreases as expected. Comparing the maximum intensity at $R = 1$ m and $R = 100$ m, the magnitude is found to have decreased 100 times.

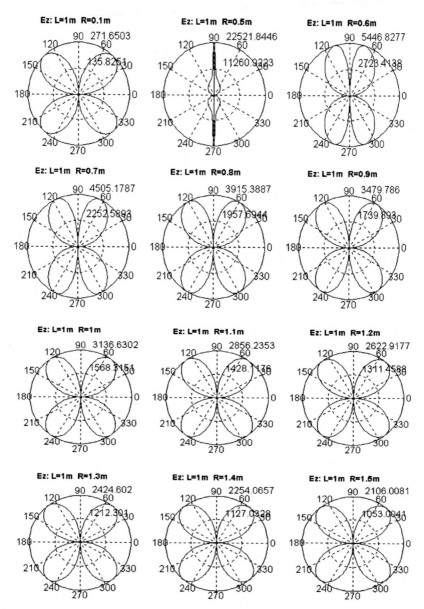

Figure 3.29: Variation of the E_z component with distance R from the center of the line.

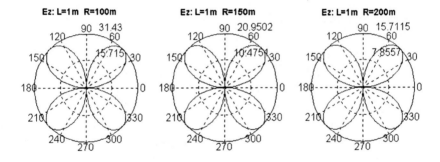

Figure 3.30: Variation of E_z at larger distance R.

The directivity of the lobes is observed to be shifting away from angular positions 90° and 270° as the distance R is increased.

Variation of the E_r component with distance R. With reference to Fig. 3.28, the distance R, that is the radius from the center of the line, is varied from 0.4 to 0.9 m. The polar plots of electric field are shown in Fig. 3.31.

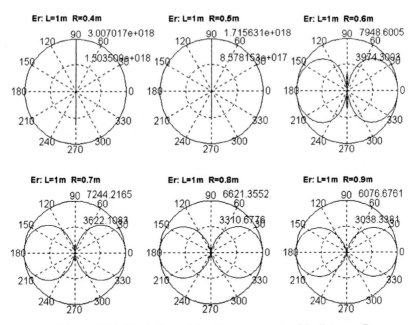

Figure 3.31: Variation of the E_r component with distance R.

The plot of E_z shows that the far field has the maximum intensity at angular positions of 0° and 90°, when the distance R is greater than half the length of the element. These are the positions orthogonal to the line element. The intensities are observed to decrease as the angular position moves toward the direction of the line element. At angular positions of 90° and 270°, the intensity is zero.

However, when the distance R is less than half the length of the line, the intensity has the highest magnitude at angular positions of 90° and 270°. This can be verified from eqn (108) in that the increase in intensity is contributed by the [I] part of the far field. At 90° and 270°, the coefficient of [I] is infinity, hence the first two plots in Fig. 3.31 just show a sharp peak. To illustrate this, the far field is plotted with the angular position varying from 0° to 89° for a fixed distance R which is less than half the length of the finite element. This is shown in Fig. 3.32. This also explains why there occurs a sidelobe in the plot at $R = 0.6$ and 0.7 m. At larger distances, the contribution of [I] is small, and hence no sidelobe is observed.

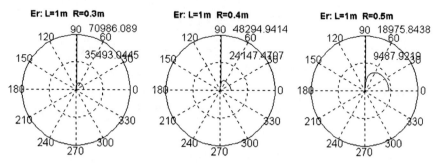

Figure 3.32: Plot showing the field at distance $R \leq L/2$.

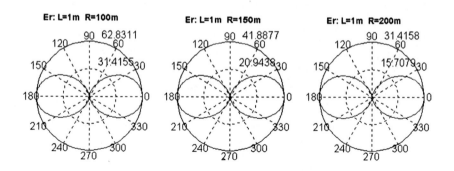

Figure 3.33: Variation of E_r at large distance R.

Variation of the resultant electric field E with distance R. With reference to Fig. 3.28, the distance R, that is the radius from the center of the line, is varied from 0.4 to 0.9 m. The polar plots of the electric field are given in Fig. 3.33.

The plots of the total electric field show that the contribution of the far-field component, E_r, is more significant than the contribution of the near-field component, E_z. The effect of near field, E_z, appears significant only at a very close distance, typically when R is less than 0.9 m. Generally, the pattern of the intensity of the E takes the same shape as the pattern of the far field. The maximum occurs at both angular positions 0° and 180°, and it diminishes as the angle moves toward the direction of the line element that is at 90° and 270°. The effect of the near field tends to broaden the lobes of the pattern.

At a very close distance R of 0.5 and 0.6 m, the characteristics of E_z were observed. This is shown in Fig. 3.34. At a larger distance, the E_r component dominates. Comparing Figs 3.35 and 3.36, the effect of the near-field portion of E_z is negligible.

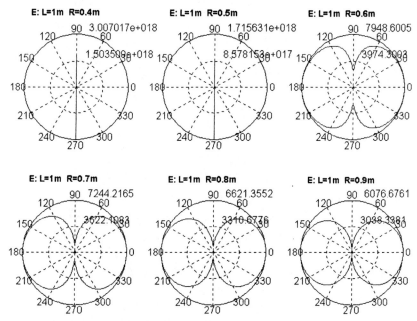

Figure 3.34: Variation of total electric field E with distance R.

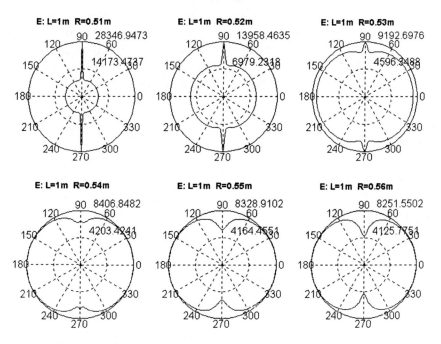

Figure 3.35: Effect of the E_z component at close distance R.

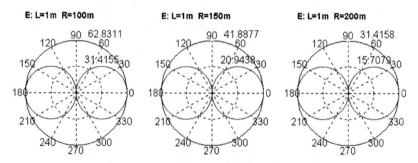

Figure 3.36: Variation of E at large distance R.

From the polar plot of the three distinct electric fields available in picocell wireless systems and imaging systems, we make the following observations:

(1) The intensity of the electric field at fixed distance R varies with angular position along the locus of R. Moreover, the pattern for both the far-field and near-field components is different. For the near-field component, the maximum intensity occurs at each quadrant of the locus circle as shown in Fig. 3.29. For the far-field component, only two regions of maximum intensity are observed along the perpendicular bisector of the finite length line element.

(2) The effect of the E_z component is dominant when the distance R is in the near-field region. The contribution of the E_r component is still apparent in the near-field region. The E_z component broadens the lobes of the intensity pattern of the E_r component as shown in Fig. 3.34. However, at larger distance R, the contribution of E_z is negligible, and the intensity pattern approximates that of the E_r component.

(3) Based on polar plots of the total electric field pattern of Fig. 3.31, the radiated waves from the finite line element cannot be approximated to a plane wave at small values of R, because of the variation of the electric field with angle. However, at larger distances, the propagation of waves can be approximated to plane waves, provided the region covered by the radiation is small.

Acknowledgement

Section 3.9 is based on Hoole and Hoole (1996) and Chan and Ngo (1999).

Appendix 3.1   Integrals used in Section 3.9.1

- $$\int_{a_1}^{a_2} \frac{a}{\left(r^2 + a^2\right)^n} \, da = -\frac{1}{2(n-1)} \left[\frac{1}{\left(r^2 + a^2\right)^{n-1}} \right]_{a_1}^{a_2}.$$

- $\displaystyle\int_{a_1}^{a_2} \frac{1}{\left(r^2+a^2\right)^n}\,da = \left[\frac{a}{2(n-1)r^2\left(a^2+r^2\right)^{n-1}} + \frac{2n-3}{2(n-1)r^2}\int_{a_1}^{a_2}\frac{da}{\left(a^2+r^2\right)^{n-1}}\right].$

$n=1.5$

$$-\int_{a_1}^{a_2} \frac{1}{\left(r^2+a^2\right)^{1.5}}\,da = -\left[\frac{a}{r^2\sqrt{a^2+r^2}}\right]_{a_1}^{a_2}.$$

$n=2$

$$-\int_{a_1}^{a_2} \frac{1}{\left(r^2+a^2\right)^{2}}\,da = -\left[\frac{a}{2r^2\left(a^2+r^2\right)} + \frac{1}{2r^2}\int_{a_1}^{a_2}\frac{da}{\left(a^2+r^2\right)}\right]_{a_1}^{a_2}$$

$$= -\left[\frac{a}{2r^2\left(a^2+r^2\right)} + \frac{1}{2r^3}\tan^{-1}\left(\frac{a}{r}\right)\right]_{a_1}^{a_2}.$$

$n=2.5$

$$-\int_{a_1}^{a_2} \frac{1}{\left(r^2+a^2\right)^{2.5}}\,da = -\left[\frac{a}{3r^2\left(a^2+r^2\right)^{1.5}} + \frac{2}{3r^2}\int_{a_1}^{a_2}\frac{da}{\left(a^2+r^2\right)^{1.5}}\right]_{a_1}^{a_2}$$

$$= -\left[\frac{a}{3r^2\left(a^2+r^2\right)^{1.5}} + \frac{2}{3r^4}\frac{a}{\sqrt{a^2+r^2}}\right]_{a_1}^{a_2}.$$

- $\displaystyle -\int_{a_1}^{a_2} \frac{a^2}{\left(r^2+a^2\right)^n}\,da$

$n=2$

$$-\int_{a_1}^{a_2} \frac{a^2}{\left(r^2+a^2\right)^{2}}\,da = \frac{1}{2}\left[\frac{a}{r^2+a^2}\right]_{a_1}^{a_2} - \frac{1}{2}\int_{a_1}^{a_2}\frac{da}{r^2+a^2}$$

$$= \frac{1}{2}\left[\frac{a}{r^2+a^2} - \frac{1}{r}\tan^{-1}\left(\frac{a}{r}\right)\right]_{a_1}^{a_2}.$$

$n = 2.5$

$$-\int_{a_1}^{a_2} \frac{a^2}{\left(r^2+a^2\right)^{2.5}}\, da = \frac{1}{3}\left[\frac{a}{\left(r^2+a^2\right)^{1.5}}\right]_{a_1}^{a_2} - \frac{1}{3}\int_{a_1}^{a_2} \frac{da}{\left(r^2+a^2\right)^{1.5}}$$

$$= \frac{1}{3}\left[\frac{a}{\left(r^2+a^2\right)^{1.5}} - \frac{a}{r^2\sqrt{a^2+r^2}}\right]_{a_1}^{a_2}$$

4 Antenna beamforming

P.R.P. Hoole

4.1 Introduction

The following issues are addressed in this chapter:

(i) Given a wire antenna, how may the current waveform imposed on the antenna be structured to obtain a desired radiation pattern? In this case the radiation pattern of the antenna is given and the current on the wire antenna is to be determined. This is termed the inverse problem in some literature. In practically implementing such an antenna, the continuous current waveform obtained may be made up of a number of small antenna elements placed in a straight line, with each element having different current magnitudes to produce the required current waveform.

(ii) Given an array antenna, how may the currents supplied to each element of the array need to be controlled to obtain a given radiation pattern or array factor? In general, the phase difference and distance between each adjacent element will be fixed; i.e. δ and d will be fixed, The problem is to obtain the current I in each element.

(iii) Adaptive antennas: how may the above two design solutions be automated, such that the beam or radiation pattern of the antenna may be automatically varied to, for instance, receive the maximum power from a mobile transmitter. In this case the magnitudes of current I and the electronic phase angle δ must be automatically controlled to steer the main beam to be always pointing at the transmitter, and for the nulls to be created in directions from which there are interfering signals (e.g. jammers in military applications) or noise signals are arriving at the antenna elements.

We consider each one of these issues in turn, and in later chapters of this book we illustrate the implementation of a technique (a) to estimate the state (e.g. position and direction of movement) of a transmitter (Chapter 7), and (b) which achieves beamforming to keep track of a cluster of mobile stations in land or airborne wireless communications (Chapter 8). Section 4.2 of this chapter has been based on Balanis (1997), and Section 4.3 on Compton (1988).

4.2 Antenna synthesis

Whereas the focus of the first three chapters of this book has been on antenna analysis and forward design, this chapter considers the inverse solution for antennas, or antenna synthesis. The analysis problem is one of determining the radiation pattern and impedance of a given antenna structure. Antenna design is the determination of the hardware characteristics (e.g. lengths, antenna geometry, currents, etc.) of a specific antenna to produce a desired radiation pattern and/or gain. Antenna synthesis is similar to antenna design and, in fact, the terms are frequently used interchangeably. However, antenna synthesis, in its broadest sense, is an *inverse problem*. In antenna synthesis or beamforming we first specify the required radiation pattern and then use a systematic method or combination of methods to arrive at an antenna configuration and also its geometrical dimensions and current excitation distribution which produce a pattern that acceptably approximates the prescribed radiation pattern.

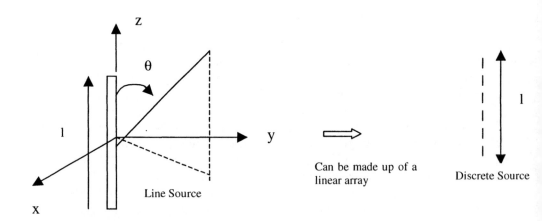

Figure 4.1: Antenna made of a continuous line-source or an array of discrete sources.

4.2.1   Line source

An antenna like a wire antenna is a continuous source in that the current flowing along the conductor is continuous along the wire. Such a continuous source may represent a true continuous source antenna like the half-wave dipole antenna, or approximate a discrete source antenna like an array antenna.

Consider the line source of length l placed along the z-axis as shown in Fig. 4.1. This may represent an array of discrete sources that have an interelement distance much smaller than the total size or length of the array. The array factor of such a continuous distribution of currents in discrete elements is sometimes called the *space factor*. The space factor (AF$_S$) of the line source shown in Fig. 4.1 is given by

$$\mathrm{AF_S}(\theta) = \int_{-l/2}^{+l/2} I_n(z')e^{j(kz'\cos\phi + \delta_n(z'))}dz',\tag{1}$$

where $I_n(z')$ and $\delta_n(z')$ are the current amplitude and phase distributions along the antenna.

For a constant phase distribution, we have $\delta_n(z') = 0$. The $kz'\cos\phi$ term comes into the equation from the array antenna factor $kd\cos\phi$ (see eqn (15) of Chapter 3), where we have divided the continuous source antenna of Fig. 4.1 into an array of discrete elements. In general, allowing for the current pulse to travel along the antenna (traveling wave antenna), $\delta_n(z') = k_z z'$; therefore the current along the antenna is given by

$$I(z') = I \exp(-k_z z').\tag{2}$$

The current travels along the antenna at a phase velocity of ω/k_z, where k_z is the current phase constant of the source. The radiation field of such a traveling wave antenna is obtained by dividing the antenna into little dipole segments and then summing up (integrating) the radiation fields from each small dipole segment. The radiation electric field of a traveling wave antenna of length L is given by

$$\mathbf{E}_\theta = \mathbf{u}_\theta jkI\eta Le^{-jkr}((\sin\theta/4\pi r)(\sin((kL/2)(\cos\theta - k_z/k)))/((kL/2)(\cos\theta - k_z/k)).\tag{3}$$

Interestingly the radiation pattern of the traveling wave antenna is always zero along the line (i.e z-axis, $\theta = 0$) and symmetrical about the axis of the antenna (z-axis). In our discussion of antenna synthesis, we shall mostly assume that the current does not propagate along the antenna, i.e. $\delta_n(z') = 0$, and hence it is a standing wave current pattern that we see on the antenna. However, for traveling wave antenna synthesis, the $\delta_n(z') = k_z z'$ phase delay should be included.

4.2.2   Fourier transform method

Given a complete description of the required pattern, this method can be used to determine the excitation of a continuous or a discrete source antenna system which yields, either exactly or approximately, the required antenna pattern. This method is commonly referred to as *beam shaping*.

4.2.2.1   Line source
Using eqn (2) we may rewrite eqn (1) for the normalized space factor as follows:

$$\begin{aligned}AF_S(\theta) &= \int_{-l/2}^{+l/2} I(z')e^{j(k\cos\phi - k_z)z'}\,dz'\\[4pt] &= \int_{-l/2}^{+l/2} I(z')e^{j\beta z'}\,dz',\end{aligned}\tag{4}$$

where

$$\beta = k\cos\phi - k_z \quad\text{or}\quad \phi = \cos^{-1}\!\left(\frac{\beta + k_z}{k}\right).\tag{5}$$

For uniform current distribution along the line, we have $I(z') = I_0/l$, and eqn (4) reduces to

$$AF_S(\phi) = I_0\,\frac{\sin\left[(kl/2)(\cos\phi - k_z/k)\right]}{(kl/2)(\cos\phi - k_z/k)}.\tag{6}$$

For observation angle ϕ to have real values, the following conditions must be satisfied:

For observation angle ϕ to have real values, the following conditions must be satisfied:

$$-(k + k_z) \leq \beta \leq (k - k_z). \tag{7}$$

The antenna is of finite length l, and beyond this length the current distribution $I(z')$ is obviously zero. Thus the limits of the integral in eqn (4) may be extended to infinity without losing the accuracy of our formulation:

$$AF_S''(\phi) = AF_S(\beta) = \int_{-\infty}^{+\infty} I(z') e^{j\beta z'} \, dz'. \tag{8}$$

Equation (8) is recognized as a Fourier transform in the spatial domain (SDFT). The corresponding inverse Fourier transform of eqn (8) is

$$I(z') = \frac{1}{2\pi} \int_{-\infty}^{+\infty} AF_S(\beta) e^{-jz'\beta} \, d\beta$$

$$= \frac{1}{2\pi} \int_{-\infty}^{+\infty} AF_S(\phi) e^{-jz'\beta} \, d\beta. \tag{9}$$

Equation (9) is the key equation for a synthesis procedure. We note that eqn (9) indicates that if $SF(\theta)$ represents the required pattern, the excitation distribution $I(z')$ that will yield the exact required pattern must, in general, exist for all values of z'. With a finite length antenna, we get an approximate $I_a(z')$ from

$$I_a(z') = \begin{cases} \dfrac{1}{2\pi} \int AF_S(\beta) e^{-jz'\beta} \, d\beta, & -l/2 \leq z' \leq l/2, \\ 0 & \text{elsewhere.} \end{cases} \tag{10}$$

From eqn (10) a very direct, but only approximate, solution for $I_a(z')$ can be obtained by using the truncated excitation distribution. Once we get $I_a(z')$, to double check how accurate our synthesis has been, we may obtain the approximate space factor resulting from this approximate solution from the following integration:

$$\text{Approx}[AF_S(\phi)] \approx \int_{-l/2}^{l/2} I_a(z') e^{j\beta z'} \, dz'. \tag{11}$$

The synthesized approximate pattern $\text{Approx}[AF_S(\theta)]$ yields the least mean square error from the specified or desired pattern $AF_S(\theta)$ over all values of β. When the values of β are restricted only in the visible region, however, the synthesized pattern will be further distorted.

Example 1. For a desired H-plane radiation pattern which is symmetrical about $\phi = \pi/2$, determine the current distribution and the approximate radiation pattern of a line source placed along the z-axis. The desired space factor is given by

$$\mathrm{AF_S}(\phi)=\begin{cases} 1, & \pi/3\leq\phi\leq2\pi/3, \\ 0 & \text{elsewhere.} \end{cases} \tag{12}$$

This is a sectoral pattern and such patterns are popular for search applications where vehicles are located by establishing communications or by a radar echo in the sector of space occupied by the main beam of the antenna pattern.

Since the pattern is symmetrical, $k_z = 0$. Using eqns (7) and (5), the values of β are given by $-k/2\leq\beta\leq k/2$ and the current distribution can be determined by eqn (9):

$$I(z')=\frac{1}{2\pi}\int_{-\infty}^{+\infty} \mathrm{AF_S}(\beta)e^{-jz'\beta}\,d\beta$$

$$=\frac{1}{2\pi}\int_{-k/2}^{k/2} e^{-jz'\beta}\,d\beta=\frac{k}{2\pi}\left[\frac{\sin(kz'/2)}{(kz'/2)}\right]. \tag{13}$$

Although we have solved for a fictitious source that exists in the limits $-\infty \leq z' \leq \infty$, a realistic approximation of the current distribution over the finite length of the line source may be written as

$$I_a(z') \approx I(z'), \quad -l/2 \leq z' \leq l/2. \tag{14}$$

By limiting the length of the line to l we will only get an approximate array factor $\mathrm{AF_S}$, which will be somewhat different to the specified or desired radiation pattern; the longer the length of the line, the closer the approximate pattern will be to the desired pattern. The approximate pattern obtained from eqn (11) using the above truncated current distribution of eqn (14) is given by

$$\mathrm{Approx}[\mathrm{AF_S}(\phi)_a]= \int_{-l/2}^{l/2} I_a(z')e^{j\beta z'}\,dz'$$

$$=\frac{1}{\pi}\left\{\int_0^{X_1}\frac{\sin x}{x}\,dx - \int_0^{X_2}\frac{\sin x}{x}\,dx\right\}, \tag{15}$$

where

$$X_1 = \frac{l}{\lambda}\pi\left(\cos\phi+\tfrac{1}{2}\right), \tag{16}$$

$$X_2 = \frac{l}{\lambda}\pi\left(\cos\phi-\tfrac{1}{2}\right). \tag{17}$$

The approximate current distribution of eqn (13) is plotted in Fig. 4.2(a). This synthesized sector pattern of eqn (15) is plotted in Fig. 4.2(b) for $l = 5\lambda$. The pattern is plotted in linear form, and in decibels, to emphasize the details of the main beam.

Observe the oscillations about the desired pattern on the main beam, called ripple, and the non-zero sidelobes. This appearance of main beam ripple and finite sidelobes is typical of any synthesized pattern.

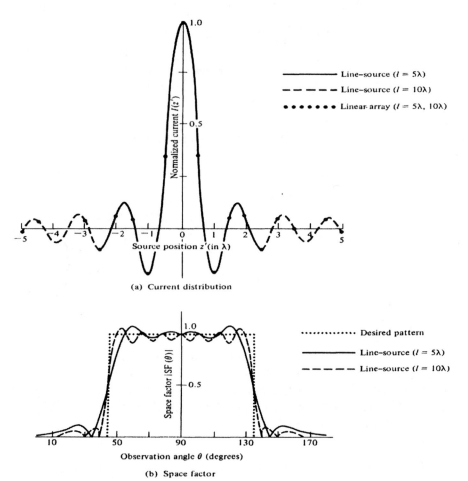

(a) Current distribution

(b) Space factor

Figure 4.2: Normalized current distribution, desired pattern, and synthesized patterns using the Fourier transform method. (From Balanis (1997). Used with permission.)

4.2.2.2 Linear array

The array factor resulting from an array of identical discrete radiators (elements) is, of course, the sum over the currents for each element weighted by the spatial phase delay from each element to the far-field point. The basic theory of an N-element array was described in Section 3.3. Consider now a linear N-element array antenna, with phase angles $-M\delta$, $-(M-1)\delta$, ..., $-\delta$, 0, δ, ..., $(M-1)\delta$, $M\delta$. The currents in the elements are not equal and are given by a_{-M}, ..., a_{-1}, a_0, a_1, ..., a_M. Since the reference point is taken at the physical center of the array, the array factor for an odd number of elements ($N = 2M + 1$) can be written as (see Section 3.3)

$$\text{AF}(\phi) = \text{AF}(\psi) = \sum_{m=-M}^{M} a_m e^{jm\psi}, \tag{18}$$

where, considering only the H-plane ($\theta = \pi/2$), we have

$$\psi = kd \cos \phi + \delta. \tag{19}$$

Note that ϕ is the angle measured away from the axis on which the array is placed; i.e. it is the angle measured from the x-axis if the array is placed along the x-axis or it is the angle measured from the z-axis if the array is placed along the z-axis, as in the present case. The elements are placed at positions along the z-axis at

$$z'_m = md, \quad m = 0, \pm1, \pm2, \ldots, \pm M. \tag{20}$$

In general, the array factor of an antenna is a periodic function of ψ, and it must repeat for every 2π radians. To satisfy periodicity requirements for real values of ϕ, we have $2kd = 2\pi$ or $d = \lambda/2$. The excitation coefficients can be determined by the Fourier formula:

$$a_m = \frac{1}{T} \int_{-T/2}^{T/2} \text{AF}(\psi)e^{-jm\psi} \, d\psi = \frac{1}{2\pi} \int_{-\pi}^{+\pi} \text{AF}(\psi)e^{-jm\psi} \, d\psi, \quad -M \le m \le M. \tag{21}$$

The Fourier series synthesis procedure is, then, to use the excitation coefficients a_m calculated from the desired pattern $\text{AF}(\phi)$ to determine the approximate array factor. This Fourier series synthesized pattern provides the least mean squared error over the region

$$\cos^{-1}\left(-2d/\lambda\right) \le \phi \le \cos^{-1}\left(2d/\lambda\right).$$

Example 2. We shall repeat the problem specified in Example 1 using an array antenna to synthesize the radiation pattern. Thus we need to determine the excitation for a broadside array whose array factor closely approximates the desired pattern. With an interelement spacing of $d = \lambda/2$ the design may be done for 11 elements and repeated for 5 elements. It is obvious that the 21-element array will give us a better radiation pattern than the 5-element array, but the cost of the antenna will increase with the number of elements.

For a broadside array, the progressive phase shift between the elements (δ) is zero. Since the pattern is non-zero only for $\pi/4 \le \phi \le 3\pi/4$, the corresponding values of ψ are obtained from eqn (19) or $-\pi/2 \le \psi \le \pi/2$. The excitation coefficients are obtained from eqn (21) or

$$a_m = \frac{1}{2\pi} \int_{-\pi/2}^{\pi/2} e^{-jm\psi} \, d\psi = \frac{1}{2}\left[\frac{\sin(m\pi/2)}{m\pi/2} \right], \tag{22}$$

and the excitation coefficients are symmetrical about the physical center (at $z = 0$) of the array (i.e. $a_m(-z'_m) = a_m(z'_m)$). The excitation coefficients, remember, indicate the relative strength of the currents that must flow in each element of the array antenna to obtain the desired radiation pattern. In general, the synthesized currents or current distribution will be real and symmetric if the desired pattern is real and symmetric, i.e. if $AF(-\phi) = AF(\phi)$; in turn, the synthesized pattern will be real and symmetric.

For $N = 21$, $d = \lambda/2$, and $l = 10\lambda$, the maximum values of the excitation coefficients (i.e. leaving out the 2^{-1} factor) are as follows:

$$a_0 = 1.0,\ a_{\pm1} = 0.3582,\ a_{\pm2} = -0.217,\ a_{\pm3} = 0.0558,\ ...,\ a_{\pm10} = -0.0100.$$

Note that, at the element positions, the line-source and linear array excitation values are identical since the two antennas are of the same length (for $N = 11$, $d = \lambda/2 \Rightarrow l = 5\lambda$).

For $N = 5$, $d = \lambda/2$ and $\psi = kd \cos\phi + \delta = \pi \cos\phi$,

$$AF(\psi) = \begin{cases} 1, & -0.5\pi < \psi < 0.5\pi, \\ 0, & -\pi < \psi < 0.5\pi,\ 0.5\pi < \psi < \pi. \end{cases} \tag{23}$$

The excitation currents should have the following ratio to achieve this array pattern: $-0.217{:}0.3582{:}1.0{:}0.3582{:}-0.217$. The root mean square quantities are as shown in Fig. 4.3.

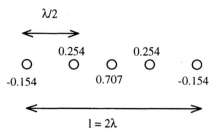

Figure 4.3: Synthesis using a five-element array. Relative amplitudes of currents are shown.

The approximate array factor that can be determined by eqn (18) will be

$$AF_a(\psi) = 0.707 + \frac{2}{\pi} \sum_{n=1}^{M} \left[\frac{1}{n} \sin(0.707 n\pi) \times \cos(n\psi) \right], \tag{24}$$

where $a_0 = 0.707$, $a_n = 1/n\pi \sin(0.707 n\pi)$.

The normalized array factors are displayed in Fig. 4.4. As expected, the larger array ($N = 21$, $d = \lambda/2$) provides a better reconstruction of the desired pattern.

4.2.3 Woodward–Lawson sampling method

A particularly convenient way to synthesize a radiation pattern is to specify values of the pattern at various points, that is, to sample the pattern. Associated with each pattern sample is a harmonic current of uniform amplitude distribution and uniform

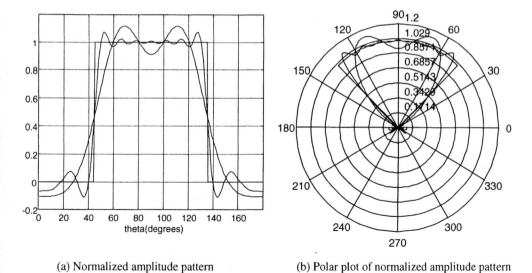

(a) Normalized amplitude pattern   (b) Polar plot of normalized amplitude pattern

Figure 4.4: Normalized desired pattern and synthesized pattern for $N = 5$ and 21 using the Fourier transform method.

progressive phase, whose corresponding field is referred to as a *composing function*. For a line source the composition function is given by

$$CF = a_m \sin(\psi_m)/\psi_m, \qquad (25)$$

and for a linear array the composition function is given by

$$CF = a_m \sin(N\delta_m)/N \sin(\delta_m). \qquad (26)$$

The excitation coefficient a_m of each harmonic current is such that the field strength is equal to the amplitude of the desired pattern at its sampled point. The total excitation of the source comprises a finite summation of space harmonics. The corresponding synthesized pattern is represented by a finite summation of composing functions, with each term representing the field of a current harmonic with uniform amplitude distribution and uniform progressive phase.

4.2.3.1 Line source

Consider a continuous current source of length l placed along the z'-axis with its center at $z' = 0$. We shall decompose this source into a set of small normalized sources. Each discrete source is assumed to have uniform amplitude and linear phase of the form

$$I_m(z') = \frac{a_m}{l} e^{-jkz' \cos \phi_m}, \quad -l/2 \le z' \le l/2. \qquad (27)$$

The angle ϕ_m represents the angles where the radiation pattern is sampled. Assuming an odd number of discrete sources, let the total current $I(z')$ be made of a sum of $2M + 1$ current elements. Each current element has the form indicated in eqn (27). Mathematically, therefore, the total current may be written as

$$I(z') = \frac{1}{l} \sum_{m=-M}^{M} a_m \, e^{-jkz' \cos \phi_m} \, , \tag{28}$$

where $m = 0, \pm1, \pm2, \ldots, \pm M$.

The array factor of each current element of eqn (28) is given by eqn (6) and may be written as

$$AF_{Sm}(\phi) = a_m \left\{ \frac{\sin\left[(kl/2)(\cos \phi - \cos \phi_m) \right]}{(kl/2)(\cos \phi - \cos \phi_m)} \right\} \tag{29}$$

The maximum value of eqn (29) occurs when $\theta = \theta_m$. Therefore the total radiation pattern of the decomposed current source is obtained by summing the $2M + 1$ radiation pattern terms each of the form given by eqn (29). Hence the approximate space factor of the continuous line source is given by

$$\text{Approx}[AF_{Sa}(\phi)] = \sum_{m=-M}^{M} a_m \left\{ \frac{\sin\left[(kl/2)(\cos \phi - \cos \phi_m) \right]}{(kl/2)(\cos \phi - \cos \phi_m)} \right\}. \tag{30}$$

While determining the values of each term on the right hand side of eqn (30), when one term becomes maximum at $\phi = \phi_m$, all other terms at the other sample points are zero at $\phi = \phi_m$. This means that at each sample point m the resultant field is equal to the single term determined at $\phi = \phi_m$. Therefore the coefficient at sample point m is given by

$$a_m = SF(\phi = \phi_m). \tag{31}$$

Thus by determining a_m at each sample point and assigning that as the value of the resultant electric field at that point, we may construct the synthesized radiation pattern. This synthesized pattern will be expected to closely resemble the desired or prescribed radiation pattern.

It should be remembered that the sample points should not be arbitrarily chosen, but the selection of the ϕ values must be such that the periodicity requirements of 2π for real values of ϕ are satisfied. Hence the sample points must be chosen such that $kz's\big|_{|z'|=l} = 2\pi$ and hence the step size is given by

$$s = \lambda/l. \tag{32}$$

By ensuring that the sample points are separated by step size s, we get electric field values at a sample point m being determined by the single coefficient a_m only. Therefore the electric field values at each point s are uncorrelated with each other when this condition is satisfied. Hence the angular location ϕ_m of each sample point must be chosen such that

$$\cos \phi_m = m \cdot s = m(\lambda/l),\qquad(33)$$

$$\text{i.e. } \phi_m = \cos^{-1}(m\lambda/l),\qquad(34)$$

with the maximum number of sample points M satisfying the condition $M \le l/\lambda$.

Example 3. Using the Woodward–Lawson synthesis method for an antenna of length $l = 5\lambda$, determine the array coefficients a_m for the problem specified in Example 1. With $l = 5\lambda$ and $M = 5$, the separation of sample points should be $s = 0.2$. Hence the total number of sampling points for odd-numbered sample points is $(M/s) + 1 = 11$. The angles where the sampling is performed are given by eqn (34):

$$\phi_m = \cos^{-1}\left(\frac{m\lambda}{l}\right) = \cos^{-1}(0.2m), \quad m = 0, \pm1, ..., \pm5.\qquad(35)$$

The angles and excitation coefficients of the sample points are given in Table 4.1.

Table 4.1: Angles and excitation coefficients of the sample points.

m	ϕ_m	$a_m = \text{SF}(\theta_m)$	m	ϕ_m	$a_m = \text{SF}(\phi_m)$
0	90°	1			
1	78.46°	1	−1	101.54°	1
2	66.42°	1	−2	113.58°	1
3	53.13°	0	−3	126.87°	0
4	36.87°	0	−4	146.13°	0
5	0°	0	−5	180°	0

4.2.3.2   Linear array

The Woodward–Lawson synthesis for a linear array follows the same line of reasoning as that given in Section 4.2.3.1 for a continuous line-source. Consider an array of length $l = (N-1)d$. Each normalized electric field sample may be written as

$$E_m(\phi) = a_m \frac{\sin\left[((N-1)/2)kd\left(\cos\phi - \cos\phi_m\right)\right]}{(N-1)\sin\left[\frac{1}{2}kd\left(\cos\phi - \cos\phi_m\right)\right]}.\qquad(36)$$

The total array factor of the entire array of individual antenna elements will be given by the summation of terms represented by eqn (36). Hence the resultant approximate array factor for the array antenna is given by

$$\text{Approx}[\text{AF}(\phi)] = \sum_{m=-M}^{M} a_m \frac{\sin\left[((N-1)/2)kd\left(\cos\phi - \cos\phi_m\right)\right]}{(N-1)\sin\left[\frac{1}{2}kd\left(\cos\phi - \cos\phi_m\right)\right]},\qquad(37)$$

where the coefficients a_m are given by

$$a_m = \text{AF}(\phi = \phi_m)\qquad(38)$$

In order to ensure that each sample is uncorrelated to neighboring samples, we must sample at sample angles ϕ_m defined by

$$\cos \phi_m = m\frac{\lambda}{l} = \frac{m\lambda}{(N-1)d} \quad \Rightarrow \quad \phi_m = \cos^{-1}\left[\frac{m\lambda}{(N-1)d}\right]. \tag{39}$$

Therefore at each element of the odd-numbered or even-numbered array, we must have the following normalized coefficient for currents (or voltages) in order to get the desired (or specified) radiation pattern:

$$a_n(z') = \frac{1}{N}\sum a_m \exp(-jkz'_n \cos \phi_m). \tag{40}$$

In eqn (40) the z-axis distance z'_n is defined as the distance of the nth element measured from the geometrical center of the array. The center element of an odd-numbered array is placed at the geometrical center of the array.

Program 4.1 contained on disk gives a listing of the computer programs in MATLAB™ for antenna synthesis.

4.3   Adaptive arrays

In the 1960s, an adaptive antenna was thought of as a self-cohering system which automatically trained its main lobe towards a desired signal of unknown direction. Such systems are easily decoyed by jammers and in any case offer no interference reduction beyond a normal beam pattern. A more recent trend was the development of null-steering systems which automatically steer pattern zeros to jammers and a lobe to the desired signal. The same technique may be used, for instance, in wireless mobile communication systems where a smart antenna constantly keeps adjusting its radiation pattern to get rid of multipath interference. Thus the basic feature of the adaptive array filter may be described as the ability of the array antenna to change (or adapt) its beam pattern according to the electromagnetic environment surrounding the antenna. The adaptation is done to satisfy a specified optimization criterion which will alter the magnitude and phase of the signal associated with each element of the array. The adaptive algorithm changes element current magnitude and phase to obtain a desired signal from the actual signal impinging on the antenna elements. To do this a third reference signal may be required. The desired signal may be acquired and its level maintained by a pilot scheme in which a reference signal is compared with the antenna output and the mean square difference is minimized. It maximizes the received power from a signal, of unknown direction, by altering their pattern while receiving by a feedback control process. If interfering signal (e.g. a jammer) or environmental noise are present, adaptive arrays will maximize $S/(N_i + N_n)$. Another advantage is that the array elements need not be linear or planar.

4.3.1 LMS adaptive array

As shown in Fig. 4.5, the quadrature hybrid splits the signal into an in-phase signal $X_I(t)$ and a quadrature signal $X_Q(t)$, also denoted here as S_{r1} and S_{r2}:

$$S_r = \frac{S_r}{\sqrt{2}} - j\frac{S_r}{\sqrt{2}} = S_{r1} + S_{r2}, \tag{41}$$

$$S_{r1} = \frac{S_r}{\sqrt{2}}, \quad S_{r2} = \frac{S_r}{\sqrt{2}} \text{ retarded by } 90°. \tag{42}$$

Hence the received signal for each element is given by

$$V_1 = V_{r1} + V_{i1} = \frac{S_1}{\sqrt{2}}(W_{r1} - jW_{i1}). \tag{43}$$

The output of each element is changed in amplitude and phase by multiplying it with the weights and so an antenna pattern to meet desired criteria may be formed.

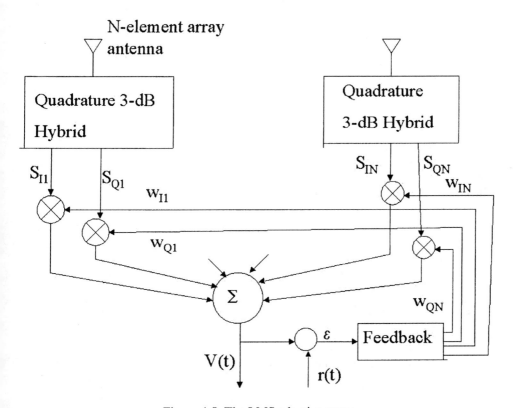

Figure 4.5: The LMS adaptive array.

The LMS array is based on a minimum mean square error concept. An error signal $\varepsilon(t)$ is obtained by subtracting the array output $s(t)$, also known as $V(t)$, from another signal called the reference signal $r(t)$.

$$\varepsilon(t) = r(t) - V(t), \tag{44}$$

where

$$V(t) = V_d(t) + n(t) + V_i(t). \tag{45}$$

Here V_d = desired signal, n = noise signal and V_i = interfering signal.
Hence the error may be written as

$$\varepsilon(t) = r(t) - V_d(t) - n(t) - V_i(t). \tag{46}$$

The mean square error (MSE) is

$$E[\varepsilon^2(t)] = E\left\{[r(t) - V_d(t)]^2\right\} + E[n^2(t)] + E[V_i^2(t)]. \tag{47}$$

The desired signal, interfering signal and the noise are all assumed to be uncorrelated zero-mean processes so that all the cross-product terms will be zero. As can be seen from eqn (47), $E[\varepsilon^2(t)]$ will be minimum only when both $E[n^2(t)]$ and $E[V_i^2(t)]$ are minimum. And the LMS array is able to make $E[V_i^2(t)]$ small by forming a pattern with nulls in the direction of V_i and n sources, while at the same time maximizing V_d by establishing the main beam in that direction. Hence when $V_d(t) \Rightarrow r(t)$ in amplitude and phase, then $E\{[r(t) - V_d(t)]^2\}$ will be small (see Fig. 4.6).

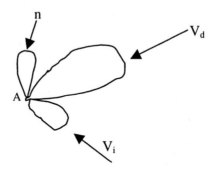

Figure 4.6: An imperfect antenna beam resulting from synthesis.

4.3.2   Two-element array

We shall first illustrate the inherent strength in using weights to adapt the antenna beam. Consider the two-element array antenna shown in Fig. 4.7. The distance between the two elements A_1 and A_2 is d (e.g. $d = \lambda/2$) and, for simplicity, we shall assume that the elements have an isotropic radiation pattern. Thus the radiation pattern of the antenna is equal to the array factor of the antenna. Consider now a jamming or interfering signal impinging on the array antenna from a direction θ_i. We do not want the antenna to receive signals arriving from the direction θ_i. Thus the resultant signal V_0 should be equal to zero for $\theta = \theta_i$. In Fig. 4.7 we have V_i = interference or jamming signal, W_2 = complex weight of the second element A_2, and V_0 = sum of the output signals.

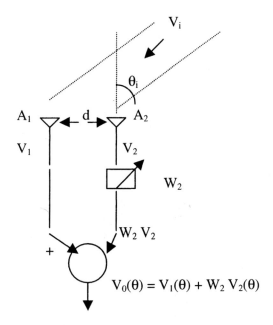

Figure 4.7: A two-element array for beamforming.

To cancel out the signal arriving from the direction θ_i we must find the value to which the complex weight should be set to, where the complex weight $W_2 = W_0 \exp(-j\delta)$. We must determine the values for W_0 and δ that will result in $V_0(\theta_i) = 0$. Let

$$V_1 = V_i e^{-jkR_1}, \tag{48}$$
$$V_2 = V_i e^{-jkR_2}, \tag{49}$$

where R_2 and R_1 are the distances from elements A_2 and A_1 to a far away transmitter, respectively. These two distances are related by

$$R_2 \approx R_1 - d \sin \theta. \tag{50}$$

We can define a signal vector X given by

$$X(\theta) = V_i \begin{bmatrix} 1 \\ e^{-jkd \sin \theta} \end{bmatrix}. \tag{51}$$

Therefore, the total array signal output is given by

$$V_0(\theta) = W^T X(\theta). \tag{52}$$

In order for the interference to be canceled out,

$$V_0(\theta_i) = 0 \quad \text{when } \theta = \theta_i.$$

This means we must have

$$W^T X(\theta_i) = 0. \tag{53}$$

By solving the above condition, we get a complex weight of   *it should be ?*

$$W = \begin{bmatrix} 1 \\ -e^{-jkd \sin \theta_i} \end{bmatrix}. \qquad \begin{bmatrix} 1 \\ -e^{+jkd \sin \theta_i} \end{bmatrix} \tag{54}$$

Thus to cancel out the interference signal coming in from direction θ_i we must have $W_2 = W_0 \exp(-j\delta)$, with $W_0 = 1$, and $\delta = \pi - kd \sin \theta_i$. The resultant signal received at any angle θ is given by

$$\begin{aligned} V_0(\theta) &= V_1(\theta) + W_2 V_2(\theta) \\ &= V_1(\theta)(1 + W_2 \exp(jkd \sin \theta)) \\ &= V_1(\theta) \, AF(\theta), \end{aligned} \tag{55}$$

where the array factor of the array antenna is given by

$$\begin{aligned} AF(\theta) &= 1 + W_2 \exp(jkd \sin \theta) \\ &= 1 - \exp(jkd(\sin \theta - \sin \theta_i)) \\ &= j2 \sin(kd(\sin \theta - \sin \theta_i)/2). \end{aligned} \tag{56}$$

As expected, if we sketch the array factor, we will have $AF(\theta_i) = 0$. Thus by setting the weight to the value we obtained, we have nulled the antenna beam in the θ_i direction. Just as we have been able to eliminate or cancel out one interference signal using a two-element array, we can cancel out $(N-1)$ interference signals using an N-element array antenna. Recalling that $k = 2\pi/\lambda$, and that the array factor is maximum when $\sin(kd(\sin \theta - \sin \theta_i)/2) = 1$, the main beam of the array will be in the direction

$$\theta = \sin^{-1}((\lambda/2d) + \sin \theta_i). \tag{57}$$

The same two-element array antenna of Fig. 4.7 can be used to maximize the main lobe in the direction of a desired signal. This possibility is used in Chapter 8 to propose a novel mobile phone unit using a smart antenna.

In the above example we assumed that the direction of the interference (jamming) signal is known. What if it is not known, as it will be in many cases? It is then that we have to use the LMS adaptive algorithm to be described in Section 4.3.3. In most cases the direction from which the undesirable signal is coming from is not known, and with the LMS algorithm we will seek to beamform in order to null the array factor in the direction of the interference signal. We may simply illustrate this principle by imposing the following requirement: the correlation between the resultant signal $V_0(t,\theta)$

and the signal received by any one of the array elements, say $V_2(t,\theta)$, should be zero. Thus we have

$$\text{Average}\{V_0(t,\theta)V_2^*(t,\theta)\} = 0, \tag{58}$$
$$\text{i.e. Average}\{(V_1(t,\theta) + W_2' V_2(t,\theta))V_2^*(t,\theta)\} = 0.$$

Hence

$$W_2 = -\text{Average}\{V_1(t,\theta)V_2^*(t,\theta)\}/(\text{Modulus}\{V_2(t,\theta)\})^2 \tag{59}$$

For the example considered above,

$$V_1(t,\theta) = V(t,\theta) \quad \text{and} \quad V_2(t,\theta) = V(t,\theta)\exp(jkd \sin \theta_i),$$

which yield

$$W_2 = -\text{Average}\{V_1(t,\theta)V_2^*(t,\theta)\}/(\text{Modulus}\{V_2(t,\theta)\})^2 = -\exp(jkd \sin \theta_i). \tag{60}$$

This is the same weight obtained previously where we assumed to know the angle of arrival of the interference signal.

Such adaptive beamforming requires components that will modify the magnitude of the signal by W_0 and phase shift the received signal by δ. Although analog phase shifters like switched-line phase shifters and attenuators like the Quad PIN diode attenuator have been used over many years, in recent times digital means of achieving this has found many applications. In digital beamforming the digitized signal is weighted in phase and amplitude by multiplication by a complex weight $W_0 \exp(-j\delta)$. Then the resultant weighted signal from each element of the array may be digitally summed. Digital beamforming has several advantages, including beamsteering at full clock rate, very accurate phase and amplitude correction, and sufficient flexibility to minimize power requirement in wireless communication and radar systems. To achieve accurate beamforming, the array antenna may be calibrated by injecting a test signal into each element of the array and measuring the amplitude and phase of the received signal; the weights are adjusted until the correct, expected signal is obtained. The test signal may be injected through wires connected from the source to the point just behind the antenna element. An alternative method is to use another antenna in the far field to transmit a known signal to the antenna to be calibrated.

4.3.3   The LMS weights

Going back to the general formulation for an N-element adaptive array, let us determine the optimum weights that yield minimum mean square error (MSE = $E[\varepsilon^2(t)]$), where the temporal error signal is given by

$$\varepsilon(t) = r(t) - V(t)$$
$$= r(t) - \sum_{\substack{j=1 \\ p=r,i}}^{N} W_{pj}S_{pj}(t). \tag{61}$$

Hence the MSE is given by

$$E[\varepsilon^2(t)] = E[r^2(t)] - 2\sum_{\substack{j=1 \\ p=r,i}}^{N} W_{pj}E[r(t)S_{pj}(t)] + \sum_{\substack{j=1 \\ p=r,i}}^{N}\sum_{\substack{k=1 \\ q=r,i}}^{N} W_{pj}W_{qk}E[S_{pj}(t)S_{qk}(t)]. \quad (62)$$

For convenience we shall define some column matrices. The $2N \times 1$ weight vector

$$W(t) = [W_{r1}(t) \ W_{i1}(t) \ W_{r2}(t) \ W_{i2}(t) \ ...]^T, \quad (63)$$

the $2N \times 1$ signal vector

$$S(t) = [S_{r1}(t) \ S_{i1}(t) \ S_{r2}(t) \ S_{i2}(t) \ ...]^T, \quad (64)$$

and the cross-correlation vector

$$P = E[S(t)r(t)]. \quad (65)$$

The above result may be written more conveniently by using matrix notation:

$$E\left[\varepsilon^2(t)\right] = E\left[r^2(t) - 2W^T P + W^T RW\right], \quad (66)$$

where the correlation matrix R is given by

$$R = E\left[S(t)S^T(t)\right]$$

$$= E\begin{bmatrix} S_{r1}(t)S_{r1}(t) & S_{r1}(t)S_{i1}(t) & S_{r1}(t)S_{r2}(t) & \cdots \\ S_{i1}(t)S_{r1}(t) & S_{i1}(t)S_{i1}(t) & S_{i1}(t)S_{i2}(t) & \cdots \\ S_{r2}(t)S_{r1}(t) & S_{r2}(t)S_{i1}(t) & S_{r2}(t)S_{r2}(t) & \cdots \\ \cdots & \cdots & \cdots & \cdots \\ \cdots & \cdots & \cdots & \cdots \\ \cdots & \cdots & \cdots & \cdots \end{bmatrix}, \quad (67)$$

and the gradient of the MSE in the weight domain is given by

$$\nabla E[\varepsilon^2(t)] = \begin{bmatrix} \dfrac{\partial E[\varepsilon^2(t)]}{\partial W_{r1}} \\[2mm] \dfrac{\partial E[\varepsilon^2(t)]}{\partial W_{i1}} \\[2mm] \dfrac{\partial E[\varepsilon^2(t)]}{\partial W_{r2}} \\[2mm] \cdots \\ \cdots \\ \cdots \end{bmatrix}. \quad (68)$$

Hence the gradient of eqn (66) becomes

$$\nabla E[\varepsilon^2(t)] = -2\nabla(W^T P) + \nabla(W^T R W) . \tag{69}$$

Now the first term on the right hand side of eqn (68) is

$$\nabla(W^T P) = \nabla \left[[W_{r1} \ W_{i1} \ W_{r2} \ldots] \begin{bmatrix} P_{r1} \\ P_{i1} \\ P_{r2} \\ \cdots \\ \cdots \end{bmatrix} \right]$$

$$= \begin{bmatrix} P_{r1} \\ P_{i1} \\ P_{r2} \\ \cdots \\ \cdots \end{bmatrix} = P . \tag{70}$$

In order to simplify the matrix relation, consider the following vector:

$$C = A^T B A = \sum_{j=1}^{N} \sum_{k=1}^{N} a_j a_k b_{jk} \tag{71}$$

$$= a_1 a_1 b_{11} + a_1 a_2 b_{12} + \cdots + a_1 a_N b_{1N}$$
$$+ a_2 a_1 b_{21} + a_2 a_2 b_{22} + \cdots + a_2 a_N b_{2N}$$
$$\cdots\cdots\cdots\cdots\cdots\cdots\cdots\cdots\cdots\cdots$$
$$\cdots\cdots\cdots\cdots\cdots\cdots\cdots\cdots\cdots\cdots$$
$$+ a_N a_1 b_{N1} + a_N a_2 b_{N2} + \cdots + a_N a_N b_{NN} . \tag{72}$$

Then the first differentiation of C is

$$\frac{\partial C}{\partial x_1} = 2a_1 b_{11} + a_2(b_{12} + b_{21}) + \cdots + a_N(b_{1N} + b_{N1})$$

$$= 2\sum_{j=1}^{N} b_{ij} a_j \quad \text{if } B \text{ is symmetric.}$$

Therefore

$$\nabla C = \left[\frac{\partial C}{\partial a_1} \ \frac{\partial C}{\partial a_2} \ \cdots \right] = 2 \left[\sum_{1}^{N} b_{ij} a_j \ \sum_{1}^{N} b_{2j} a_j \ \cdots \right]^T . \tag{73}$$

Hence eqn (69) reduces to

$$\nabla E\left[\varepsilon^2(t)\right] = -2P + 2RW . \tag{74}$$

The weight vector yielding minimum $E[\varepsilon^2(t)]$, which we denote by W_{opt}, may be found by setting

$$\nabla E\left[\varepsilon^2(t)\right] = 0.$$

From eqns (74) we get the optimum weight,

$$W_{opt} = R^{-1}P. \tag{75}$$

This is also the steady state weight vector.

We shall now consider how the weights are to be initialized to get the routine working. To ensure that $\varepsilon(t)$ moves towards a minimum, we must consider the time-domain differential

$$\frac{d\left(E\left[\varepsilon^2(t)\right]\right)}{dt} = \sum_{\substack{j=1 \\ p=r,i}}^{N} \frac{\partial E\left[\varepsilon^2(t)\right]}{\partial W_{pj}} \cdot \frac{dW_{pj}}{dt}. \tag{76}$$

As shown in Fig. 4.8, the weights converge to the desired weights when the slope of the curve becomes zero.

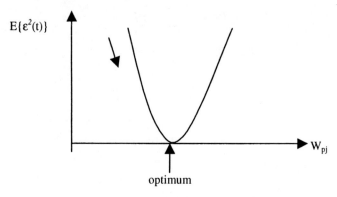

Figure 4.8: Defining the optimum weights.

For $E[\varepsilon^2(t)]$ to decrease with time, the derivative of it has to be negative, which means

$$\frac{dW_{pj}}{dt} = -\alpha \frac{\partial E[\varepsilon^2(t)]}{\partial W_{pj}}, \tag{77}$$

where α is a positive real constant. Thus

$$\frac{dW}{dt} = -\alpha \nabla\left[E\left(\varepsilon^2(t)\right)\right], \tag{78}$$

from which we get

$$\Rightarrow \qquad \frac{d\left[E\left(\varepsilon^2(t)\right)\right]}{dt} = -\alpha \sum_{\substack{j=1 \\ p=r,i}}^{N} \left\{\frac{\partial E\left[\varepsilon^2(t)\right]}{\partial W_{pj}}\right\}^2 \tag{79}$$

Equation (79) yields a value which is always negative. Hence,

$$\frac{\partial E\left[\varepsilon^2(t)\right]}{\partial W_{pj}} = -\frac{\partial}{\partial W_{pj}}\left\{2\sum_{\substack{j=1 \\ p=r,i}}^{N} W_{pj} E\left[r(t)S_{pj(t)}\right]\right\} + \frac{\partial}{\partial W_{pj}}\left\{\sum_{\substack{j=1 \\ p=r,i}}^{N}\sum_{\substack{k=1 \\ q=r,i}}^{N} W_{pj}W_{qk} E\left[S_{pj}(t)S_{qk}(t)\right]\right\}$$

$$= -2E\left[r(t)S_{pj}(t)\right] + 2\sum_{\substack{k=1 \\ q=r,i}}^{N} W_{qk} E\left[S_{pj}(t)S_{qk}(t)\right]$$

$$= -2E\left\{S_{pj}(t)\left[r(t) - \sum_{\substack{k=1 \\ q=r,i}}^{N} W_{qk}S_{qk}(t)\right]\right\}$$

$$= -2E\left[S_{pj}(t)\varepsilon(t)\right]. \tag{80}$$

Hence from eqn (76) we get

$$\frac{dW_{pj}}{dt} = 2\alpha E\left[S_{pj}(t)\varepsilon(t)\right]. \tag{81}$$

Equation (81) in vector form yields

$$\frac{dW}{dt} = 2\alpha E\left[\varepsilon(t)S(t)\right]. \tag{82}$$

Since it takes time to obtain the expected value (by a time average), we can instead estimate $E\left[\varepsilon(t)S_{pj}\right]$ by $S_{pj}\varepsilon(t)$. Equation (82) yields

$$\frac{dW_{pj}}{dt} = 2\alpha S_{pj}\varepsilon(t), \tag{83}$$

which is known as the Widrow LMS algorithm.

As $\varepsilon(t)$ and S_{pj} are stochastic, W_{pj} will vary randomly about their mean values. Hence α should be chosen to be small enough so that the weights average out the stochastic variations. The implementation of the algorithm is shown in Fig. 4.9.

In the time domain, the weight is determined from

$$W_{pj}(t) = 2\alpha \int S_{pj}(t)\varepsilon(t)\,dt, \tag{84}$$

with each feedback loop forming the product $2\alpha S_{pj}(t)\varepsilon(t)$ and integrating it to give the weights. The flow chart for an adaptive antenna algorithm is given in Fig. 4.10. The direction of arrival (DOA) of the received signal is assumed to be known. The estimation of the position of a mobile transmitter, and hence the DOA, is described in Chapter 7.

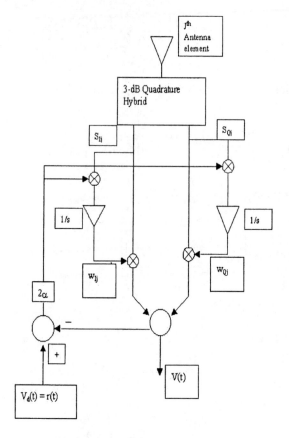

Figure 4.9: Implementation for one element.

4.3.4   Complex signal notation

One way of simplifying the analysis of an adaptive array is the use of analytic signal notation with complex weights. We redefine some column matrices as follows:

$$W(t) = [W_1(t) \;\; W_2(t) \;\; W_3(t) \;\; W_4(t) \dots]^T, \tag{85}$$
$$S(t) = [S_1(t) \;\; S_2(t) \;\; S_3(t) \;\; S_4(t) \dots]^T, \tag{86}$$
$$P = E[S(t)^* r(t)], \tag{87}$$
$$R = E[S(t)^* S(t)^T], \tag{88}$$

where W = weight factor, S = signal vector, P = correlation vector and R = covariance matrix.

In place of eqn (66), i.e. $E[\varepsilon^2(t)] = E[r^2(t) - 2W^T P + W^T RW]$, we have

$$E\left[\varepsilon^2(t)\right] = E\left[r^2(t)\right] - 2\,\mathrm{Re}\left(W^T P\right) + W^T RW . \tag{89}$$

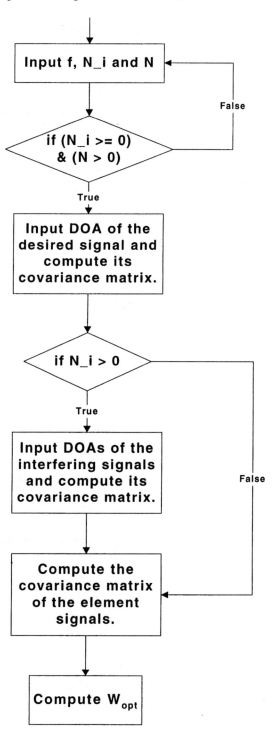

Figure 4.10: Flow chart for an adaptive antenna algorithm.

Defining $W = W_r - jW_j$ and $P = P_r - jP_j$, the equation becomes

$$E\left[\varepsilon^2(t)\right] = E\left[r^2(t)\right] - 2W_r^T P_r + 2W_j^T P_j + W^T RW. \tag{90}$$

The quadratic form $W^T RW$ is shown to have partial derivatives of

$$\partial\left(W^T RW\right)/\partial W_r^n = 2\,\text{Re}\left(RW\right)^n, \tag{91}$$

$$\partial\left(W^T RW\right)/\partial W_j^n = 2\,\text{Im}\left(RW\right)^n. \tag{92}$$

Using these results, we have

$$\partial E\left[\varepsilon^2(t)\right]/\partial W_r^n = -2P_r^n + 2\,\text{Re}\left(RW\right)^n, \tag{93}$$

$$\partial E\left[\varepsilon^2(t)\right]/\partial W_j^n = 2P_j^n + 2\,\text{Im}\left(RW\right)^n. \tag{94}$$

Equating all partial derivatives to zero yields

$$\text{Re}\left(RW\right) = P_r, \tag{95}$$

$$\text{Im}\left(RW\right) = -P_j, \tag{96}$$

and adding the previous two equations gives the final complex form

$$RW = \text{Re}\left(RW\right) + j\,\text{Im}\left(RW\right) = P_r - jP_j = P, \tag{97}$$

which yields

$$W = R^{-1}P. \tag{98}$$

This is the steady state weight vector. The analytic output signal for the entire array is

$$\tilde{s}(t) = W^T S(t), \tag{99}$$

and the voltage pattern of the array is the normalized magnitude of the output signal. A listing of a program for adaptive beamforming is given in Appendix 4.2.

Acknowledgement

We acknowledge the contribution of Mr. Timmy Tan to the development of this chapter and the MATLAB™ programs contained on disk.

5 Antennas and signals in magnetic resonance imaging

T.S. Naveendra and P.R.P. Hoole

5.1 Introduction

In order to model the radiation mechanism and the electromagnetic fields associated with the atomic nuclei, it is important to understand the physics of the nuclear resonance phenomenon in a mathematical context. The equation of motion of resonating atomic nuclei is presented mathematically and the magnetic moment in the imaging plane under consideration is reformulated as a combination of two orthogonal magnetic moments which are 90° out of phase. It is then possible to analyze the near-field electromagnetic fields due to these magnetic moments more closely using the established field equations. Following the derivation of temporal return signals in MRI, the effect of the near electromagnetic fields on the magnetic resonance images is studied and correction for the undesirable effects is proposed and implemented. In Chapter 7, the proper characterization of near-field MRI signals is used to develop an MRI diagnostic tool. This chapter is based on work reported in Naveendra and Hoole (1999).

5.2 Equation of motion of nuclear magnetic moment

An atomic nucleus has its own intrinsic angular momentum vector S, called spin, associated with a unique quantum number $I \equiv \{1/2(j-1); j = 1, 2, 3,...\}$. A nucleus with $I \neq 0$ has its own intrinsic nuclear magnetic moment m, related to S by

$$m = \gamma \cdot S, \tag{1}$$

where γ is called the gyromagnetic ratio and it depends on the type of nuclei. This allows a particular type of nucleus to be selected and its magnetic resonance characteristics to be studied.

In the presence of an external magnetic field B, **k**, the magnetic moment of the nucleus, will experience a torque T given by

$$T = m \times B, \tag{2}$$

which must in turn be equal to the rate of change of the angular momentum. Using eqns (1) and (2),

$$\frac{dm}{dt} = \gamma m \times B. \tag{3}$$

Equation (3) is the equation of motion of the magnetic moment. It could be written componentwise as follows:

$$\frac{dm_x}{dt} = \gamma\left[m_y H_z - m_z H_y\right],$$

$$\frac{dm_y}{dt} = \gamma\left[m_z H_x - m_x H_z\right], \tag{4}$$

$$\frac{dm_z}{dt} = \gamma\left[m_x H_y - m_y H_x\right].$$

5.2.1   Bloch equations

Equation (3) has been obtained ignoring the interactions of the nuclei with the surrounding molecular environment (spin–lattice interaction) and those between close nuclei (spin–spin interaction). The spin–lattice interaction involves the transfer of energy from the spin system to the molecular environment by the random thermal motion of the spins. In spin–spin interaction, the weak magnetic field

$$H_i(x,t) \propto -\nabla\left(\frac{m_i(t)[x - x_i(t)]}{|x - x_i(t)|^3}\right), \tag{5}$$

generated by the ith nucleus, influences the spins nearest to it, changing their Larmor frequency. The solution to eqn (3) without considering these interactions will be

$$m_x(t) = m_x(t_0)\cos\omega_0(t - t_0) + m_y(t_0)\sin\omega_0(t - t_0), \tag{6}$$

$$m_y(t) = m_y(t_0)\cos\omega_0(t - t_0) - m_x(t_0)\sin\omega_0(t - t_0), \tag{7}$$

$$m_z(t) = m_z(t_0), \tag{8}$$

where t_0 is the initial time and $\omega_0 = -\gamma H_0$ is the Larmor frequency. When $t \to \infty$, the solution is unbounded, which is contrary to experimental observations. Therefore eqn (3) must be modified to properly represent the macroscopic behavior of the physical system. The modifications were proposed by Bloch and are based on the following observations:

(i)   The changes in the spins' total energy density E are only due to spin–lattice interaction, while spin–spin interaction does not involve changes in E. Since the dominant term of $E(t)$ is $m(t) \cdot H_0(t)$, the spin–lattice interaction will contribute to the changes of $m_z(t)$. Thus it is clear that $m_z(t)$ approaches the equilibrium value M_0 exponentially with a time constant T_1 called longitudinal or spin–lattice relaxation time, so that the rate of change of $m(t)$ due to spin–lattice interaction will be $-[m(t) - M_0]/T_1$.

(ii) Spin–spin interaction does not influence E. Hence it will not be involved in the changes of $m_z(t)$. But it will affect $m_x(t)$ and $m_y(t)$. Therefore the transversal components $m_x(t)$ and $m_y(t)$ of $m(t)$ will decrease to zero exponentially with a time constant T_2 called transversal or spin–spin relaxation time. The rates of change of $m_x(t)$ and $m_y(t)$ are $-m_x(t)/T_2$ and $-m_y(t)/T_2$ respectively.

The effect of decay of $m_x(t)$ and $m_y(t)$ due to spin–spin relaxation is stronger when the mobility of the nuclei decreases. This is the reason why the T_2 values of solids are smaller than those of liquids. Hence the transient MRI signals from solids are reduced to zero much faster than those from liquids. In the MRI image solids do not appear or appear according to the color code denoting the least strength. Hence the skull will appear as a dark region in an MR image whereas it will appear bright in a CT scan.

In order to take into account the perturbation due to spin–lattice and spin–spin interaction, the decay rates computed are applied to eqn (4), resulting in

$$\frac{dm_x}{dt} = \gamma \left[m_y H_z - m_z H_y \right] - \frac{m_x}{T_2},$$

$$\frac{dm_y}{dt} = \gamma \left[m_z H_x - m_x H_z \right] - \frac{m_y}{T_2}, \tag{9}$$

$$\frac{dm_z}{dt} = \gamma \left[m_x H_y - m_y H_x \right] - \frac{m_z - M_0}{T_1}.$$

These equations are basic in MR experiments and called Bloch equations. If $W(x,t)$ is defined as the vector with components

$$\left(-\frac{m_x}{T_2} , \quad -\frac{m_y}{T_2} , \quad -\frac{m_z - M_0}{T_1} \right),$$

the Bloch equations in eqn (9) become

$$\frac{dm(x,t)}{dt} = \gamma m(x,t) \times H(x,t) + W(x,t), \tag{10}$$

where

$$\mathbf{H}(x,t) = H_0 \, \mathbf{k} + H_1(x,t) \, \mathbf{k}. \tag{11}$$

Bloch equations are solved analytically, with the assumption that the external magnetic field is in the z direction, which is often the case in practice. In order to solve eqn (10), a rotating orthonormal frame $\mathfrak{R}_r(t)$ is introduced such that it has the same origin as the normal frame $\mathfrak{R}(t)$. For a point $P \equiv x$, the vectors in eqn (10) can be assumed applied to the origin of $\mathfrak{R}$. Let the magnetic moment in the rotating, stationary frame be related by

$$m(t) = R(t) \, m(t)', \tag{12}$$

where the prime denotes a quantity in the rotating frame. $R_r(t)$ is an orthonormal matri whose columns are the coordinates of the vectors of the frame $\Re_r(t)$ expressed in ' frame coordinates. Since $R_r(t)$ is orthonormal,

$$R_r(t)R_r^{\mathrm{T}}(t) = I, \tag{13}$$

where I denotes the identity matrix and T denotes matrix transposition operation.

Taking the derivative on both sides of eqn (12) yields

$$\frac{d}{dt}R_r(t)R_r^{\mathrm{T}}(t) = -R_r(t)\frac{d}{dt}R_r^{\mathrm{T}}(t)$$

$$= -\left(\frac{d}{dt}R_r(t)R_r^{\mathrm{T}}(t)\right)^{\mathrm{T}}. \tag{14}$$

Therefore $(d/dt)R_r(t)R_r^{\mathrm{T}}(t)$ is an anti-symmetric matrix and a vector $\omega(x,t)=[\omega_x(x, \omega_y(x,t), \omega_z(x,t)]^{\mathrm{T}}$ can be defined such that

$$\frac{d}{dt}R_r(t)R_r^{\mathrm{T}}(t) = \begin{bmatrix} 0 & -\omega_z(x,t) & \omega_y(x,t) \\ \omega_z(x,t) & 0 & -\omega_x(x,t) \\ -\omega_y(x,t) & \omega_x(x,t) & 0 \end{bmatrix} \tag{15}$$

and

$$\left(\frac{d}{dt}R_r(t)R_r^{\mathrm{T}}(t)\right)m(x,t) = \omega(x,t) \times m(x,t). \tag{16}$$

Applying operator $R_r^{\mathrm{T}}(t)$ on both sides of eqn (10), we get

$$R_r^{\mathrm{T}}(t)\frac{d}{dt}m(x,t) = \gamma\left[R_r^{\mathrm{T}}(t)m(x,t)\right] \times \left[R_r^{\mathrm{T}}(t)H(x,t)\right] + R_r^{\mathrm{T}}(t)W(x,t)$$

$$= \gamma m'(x,t) \times H'(x,t) + W'(x,t), \tag{1}$$

where

$$m'(x,t) = R_r^{\mathrm{T}}(t)\, m(x,t),$$

$$H'(x,t) = R_r^{\mathrm{T}}(t)\, H(x,t),$$

$$W'(x,t) = R_r^{\mathrm{T}}(t)\, W(x,t).$$

Taking the time derivative on both sides of eqn (11) leads to

$$\frac{\mathrm{d}}{\mathrm{d}t}m(x,t) = \frac{\mathrm{d}}{\mathrm{d}t}R_r(t)\,m'(x,t) + R_r(t)\frac{\mathrm{d}}{\mathrm{d}t}m'(x,t)$$

$$= \left(\frac{\mathrm{d}}{\mathrm{d}t}R_r(t)\,R_r^{\mathrm{T}}(t)\right)m(x,t) + R_r(t)\frac{\mathrm{d}}{\mathrm{d}t}m'(x,t)$$

$$= \omega(x,t) \times m(x,t) + R_r(t)\frac{\mathrm{d}}{\mathrm{d}t}m'(x,t),$$

which on applying $R_r^{\mathrm{T}}(t)$ operator on both sides reduces to

$$R_r^{\mathrm{T}}(t)\frac{\mathrm{d}}{\mathrm{d}t}m(x,t) = \omega'(x,t) \times m'(x,t) + \frac{\mathrm{d}}{\mathrm{d}t}m'(x,t), \tag{18}$$

where $\omega'(x,t) = R_r^{\mathrm{T}}(t)\,\omega(x,t)$.

Equating eqns (17) and (18) yields

$$\frac{\mathrm{d}}{\mathrm{d}t}m'(x,t) - m'(x,t) \times \omega'(x,t) = \gamma m'(x,t) \times H'(x,t) + W'(x,t),$$

i.e.

$$\frac{\mathrm{d}}{\mathrm{d}t}m'(x,t) = \gamma m'(x,t) \times \left[H'(x,t) + \frac{\omega'(x,t)}{\gamma} \right] + W'(x,t). \tag{19}$$

Here, $\omega'(x,t)$ and $H'(x,t) + (\omega'(x,t)/\gamma)$ are called the angular velocity and effective magnetic field respectively. In the rotating frame,

$$\omega'(x,t) = -\gamma H'(x,t). \tag{20}$$

Hence eqn (19) becomes

$$\frac{\mathrm{d}}{\mathrm{d}t}m'(x,t) = W'(x,t), \tag{21}$$

i.e.

$$\frac{\mathrm{d}m'_x(x,t)}{\mathrm{d}t} = -\frac{m'_x(x,t)}{T_2(x)},$$

$$\frac{\mathrm{d}m'_y(x,t)}{\mathrm{d}t} = -\frac{m'_y(x,t)}{T_2(x)}, \tag{22}$$

$$\frac{\mathrm{d}m'_z(x,t)}{\mathrm{d}t} = -\frac{m'_z(x,t) - M_0}{T_1(x)}.$$

Solutions of eqn (22) are

$$m'_x(x,t) = m'_x(x,0) \, \exp\left(-\frac{t}{T_2(x)}\right),$$

$$m'_y(x,t) = m'_y(x,0) \, \exp\left(-\frac{t}{T_2(x)}\right), \tag{23}$$

$$m'_z(x,t) = m'_z(x,0) - \left[m'_z(x,0) - M_0(x)\right]\left[1 - \exp\left(-\frac{t}{T_1(x)}\right)\right],$$

where $\mathbf{m}'(x,0)$ is the initial magnetization.

To express the solutions in the stationary plane, the quantities should be rotated by an angle θ given by

$$\theta(x,t) = \int_0^t \omega(x,t') \, dt' = -\gamma \int_0^t H(x,t') \, dt'. \tag{24}$$

Using the operator $A_r(t)$, defined as

$$A_r(t) = \begin{bmatrix} \cos\theta(x,t) & -\sin\theta(x,t) & 0 \\ \sin\theta(x,t) & \cos\theta(x,t) & 0 \\ 0 & 0 & 0 \end{bmatrix}, \tag{25}$$

on eqn (23) yields the solutions in the stationary plane as

$$m_x(x,t) = \left[m_x(x,0)\cos\theta(x,t) - m_y(x,0)\sin\theta(x,t)\right]\exp\left(-\frac{t}{T_2(x)}\right), \tag{26a}$$

$$m_y(x,t) = \left[m_x(x,0)\sin\theta(x,t) + m_y(x,0)\cos\theta(x,t)\right]\exp\left(-\frac{t}{T_2(x)}\right), \tag{26b}$$

$$m_z(x,t) = m_z(x,0) - \left[m_z(x,0) - M_0(x)\right]\left[1 - \exp\left(-\frac{t}{T_1(x)}\right)\right]. \tag{26c}$$

Equations (26a) and (26b) can be written as

$$m_x(x,t) = \left|m_{xy}(x,0)\right|\exp\left(-\frac{t}{T_2(x)}\right)\cos\left[\theta(x,t) + \psi(x)\right], \tag{27a}$$

$$m_y(x,t) = \left|m_{xy}(x,0)\right|\exp\left(-\frac{t}{T_2(x)}\right)\sin\left[\theta(x,t) + \psi(x)\right], \tag{27b}$$

where $m_{xy}(x,t) \equiv \left[m_x(x,t), m_y(x,t), 0 \right]$ is the projection of $\mathbf{m}(x,t)$ on the xy-plane and $\psi(x) = m_y(x,0) / m_x(x,0)$.

From eqns (27a) and (27b), it is seen that $\mathbf{m}(x,t)$ rotates around $\mathbf{k}$ with an angular velocity of $\omega(x,t) = -\gamma H(x,t)$ (Larmor frequency) and its amplitude decreases with a time constant T_2. From eqn (26c) it is seen that the $\mathbf{k}$ projection of $\mathbf{m}(x,t)$ approaches exponentially the value $M_0(x)$ with time constant T_1.

5.3   Electromagnetic fields due to the resonating nuclear magnetic moment

In conventional magnetic resonance signal analysis, the time domain signal expressions are obtained in terms of the net magnetization moment and no study has been carried out on the fields created by these magnetic moments. In analyzing the signals, the effect of the electromagnetic fields, its interaction with the sensing coil and especially the effect of the near electromagnetic fields have not been studied.

The rotating magnetic moment in the xy-plane could be modeled by two magnetic moments which are in phase quadrature and acting along the x- and y-axis. These two magnetic moments could be represented by eqns (27a) and (27b) respectively. The geometrical setup of the magnetic moments is shown in Fig. 5.1.

In order to find the electric and magnetic fields due to the rotating magnetic moment, the rotating magnetic moment is replaced by two magnetic moments in the x, y directions which are in phase quadrature. This will create a rotating magnetic moment in the xy-plane. The fields due to magnetic moments in the x, y directions are found separately and added in the time domain vectorially to arrive at the resultant electromagnetic fields.

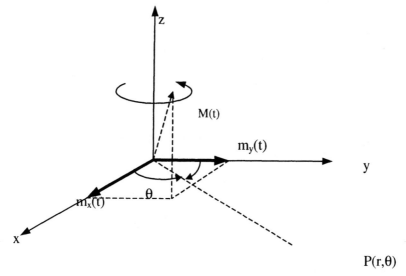

Figure 5.1: Geometrical placement of the magnetic moments.

The electromagnetic fields due to a magnetic moment were computed in Chapter 2 and could be used here. The magnetic moment m is related to the hypothetical magnetic current I_m given by

$$m = I_m l,$$ (28)

where l is the length of the magnetic dipole.

Using results obtained in Section 2.15 of Chapter 2, H_r and H_θ could be written as

$$H_r = \frac{I_m l \cos \theta}{2\pi \eta \, r^2} \left[1 + \frac{1}{jkr} \right] e^{-jkr},$$

$$H_\theta = j \frac{k I_m l \sin \theta}{4\pi \eta \, r} \left[1 + \frac{1}{jkr} - \frac{1}{(kr)^2} \right] e^{-jkr}.$$

5.3.1   Magnetic fields due to the magnetic moment along the x-axis

Substituting for magnetic moment $I_m l$ using eqn (27a), the magnetic field components due to $m_x(t)$ are found to be

$$H_{rx}(t) = \frac{m_{xy} \, e^{-t/T_2}}{2\pi \eta} \frac{\cos \theta}{r^2} \left[1 + \frac{1}{jkr} \right] \cos \left[\omega_0 t + \psi(x) - kr \right],$$ (29)

$$H_{\theta x}(t) = \frac{k m_{xy} \, e^{-t/T_2}}{4\pi \eta} \frac{\sin \theta}{r} \left[1 + \frac{1}{jkr} - \frac{1}{(kr)^2} \right] \sin \left[\omega_0 t + \psi(x) - kr \right].$$ (30)

5.3.2   Magnetic fields due to the magnetic moments along the y-axis

Substituting for magnetic moment $I_m l$ using eqn (27b), the magnetic field components due to $m_y(t)$ are found to be

$$H_{ry} = \frac{m_{xy} \, e^{-t/T_2}}{2\pi \eta} \frac{\cos(90° - \theta)}{r^2} \left[1 + \frac{1}{jkr} \right] \sin \left[\omega_0 t + \psi(x) - kr \right],$$

which reduces to

$$H_{ry} = \frac{m_{xy} \, e^{-t/T_2}}{2\pi \eta} \frac{\sin \theta}{r^2} \left[1 + \frac{1}{jkr} \right] \sin \left[\omega_0 t + \psi(x) - kr \right],$$ (31)

and the θ component of the magnetic field is given by

$$H_{\theta y}(t) = -j\frac{km_{xy}\,e^{-t/T_2}}{4\pi\eta}\,\frac{\sin(90^{\circ}-\theta)}{r}\left[1+\frac{1}{jkr}-\frac{1}{(kr)^2}\right]\sin\left[\omega_0 t + \psi(x) - kr\right],$$

which reduces to

$$H_{\theta y}(t) = -\frac{km_{xy}\,e^{-t/T_2}}{4\pi\eta}\,\frac{\cos\theta}{r}\left[1+\frac{1}{jkr}-\frac{1}{(kr)^2}\right]\cos\left[\omega_0 t + \psi(x) - kr\right]. \tag{32}$$

5.3.3 Resultant magnetic fields at P(r,θ)

The resultant magnetic field components at an observation point $P(r,\theta)$ are found by the vector sum of the individual components (Fig. 5.2), and may be expressed as

$$E_R(t) = E_{rx}(t) + E_{ry}(t), \tag{33}$$

$$E_{\theta}(t) = E_{\theta x}(t) - E_{\theta y}(t). \tag{34}$$

Using eqns (29) and (31), we get

$$H_R = \frac{m}{2\pi\eta}\,\frac{e^{-t/T_2}}{r^2}\left[1+\frac{1}{jkr}\right]\left\{\cos\theta\cos\left[\omega_0 t + \psi(x) - kr\right] + \sin\theta\sin\left[\omega_0 t + \psi(x) - kr\right]\right\},$$

which reduces to

$$H_R = \frac{m}{2\pi\eta}\,\frac{e^{-t/T_2}}{r^2}\left[1+\frac{1}{jkr}\right]\cos\left[\omega_0 t + \psi(x) - kr - \theta\right]. \tag{35}$$

Using eqns (30) and (35), we get

$$H_{\theta}(t) = \frac{km}{4\pi\eta}\,\frac{e^{-t/T_2}}{r}\left[1+\frac{1}{jkr}-\frac{1}{(kr)^2}\right]\left\{\cos\theta\cos\left[\omega_0 t + \psi(x) - kr\right]\right.$$
$$\left. -\sin\theta\sin\left[\omega_0 t + \psi(x) - kr\right]\right\},$$

which reduces to

$$H_{\theta}(t) = \frac{km}{4\pi\eta}\,\frac{e^{-t/T_2}}{r}\left[1+\frac{1}{jkr}-\frac{1}{(kr)^2}\right]\cos\left[\omega_0 t + \psi(x) - kr + \theta\right]. \tag{36}$$

The resultant magnetic field components in the radial and tangential direction on the *xy*-plane are given by eqns (35) and (36) respectively.

5.4   Magnetic resonance time domain signals

There exists a changing magnetic field near a sample excited by an external magnetic field **H** as seen from eqns (35) and (36). Thus, a coil placed near the sample with its axis on the *xy*-plane will detect a voltage induced in it by electromagnetic induction. This is the only measurable quantity in an MR experiment and image reconstruction is done using these measurements.

To find the voltage induced in a coil, it is necessary to find the flux linkage in the coil due to the magnetic field. The flux linkage ϕ_n is defined as

$$\phi_n = \mathbf{B} \cdot \mathbf{A} = \mu \mathbf{H} \cdot \mathbf{A}, \tag{37}$$

where **A** is the area of the coil and uniform flux distribution over the coil area is assumed.

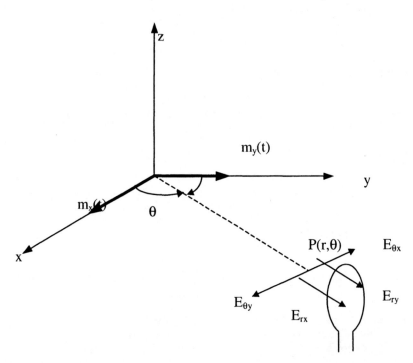

Figure 5.2: Magnetic field components due to the two magnetic moments.

When the coil is placed in a position to pick up only the radial component of the magnetic field, eqn (37) becomes

$$\phi_n = \mu H_R A, \tag{38}$$

assuming that the magnetic field H_R is uniform within the area A of the coil.

The induced voltage or current in the coil is the time derivative of the flux linkage through the coil. Hence the current is given by

$$i(t) = -C \frac{d\phi_n}{dt} \quad \text{for } t \geq t_d, \tag{39}$$

$$i(t) = -C \frac{\mu m A}{2\pi \eta} \frac{1}{r^2} \left[1 + \frac{1}{jkr} \right] \frac{d}{dt} \left\{ e^{-t/T_2} \cos \left[\omega_0 t + \psi(x) - kr - \theta \right] \right\}$$

$$= -C \frac{\mu m A}{2\pi \eta} F_R(r) \frac{d}{dt} \left\{ e^{-t/T_2} \cos \left[\omega_0 t + \psi(x) - kr - \theta + \alpha(r) \right] \right\}, \tag{40}$$

where

$$F_R(r) = \sqrt{\frac{1}{r^4} + \frac{1}{k^2 r^6}} \quad \text{and} \quad \alpha(r) = \tan^{-1} \left(-\frac{1}{kr} \right).$$

$$i(t) = -C \frac{\mu m A}{2\pi \eta} e^{-t/T_2} F_R(r)$$

$$\times \left\{ -\omega_0 \sin \left[\omega_0 t + \psi(x) - kr - \theta + \alpha(r) \right] - \frac{1}{T_2} \cos \left[\omega_0 t + \psi(x) - kr - \theta + \alpha(r) \right] \right\}.$$

$$\tag{41a}$$

Note that $\eta = \sqrt{(\mu/\varepsilon)}$, $v = 1/\sqrt{(\mu\varepsilon)}$ and wave number $k = \omega_0/v$. Substituting for these parameters, eqn (41a) becomes

$$i(t) = -C \frac{m A F_R(r)}{2\pi} e^{-t/T_2}$$

$$\times \left\{ -k \sin \left[\omega_0 t + \psi(x) - kr - \theta + \alpha(r) \right] - \frac{1}{v T_2} \cos \left[\omega_0 t + \psi(x) - kr - \theta + \alpha(r) \right] \right\}.$$

$$\tag{41b}$$

Here v is in the order of 10^8 and T_2 in the order of 10^{-3} s. In comparison to the first term of eqn (41b), the second term will be negligibly small. Thus eqn (41b) simplifies to

$$i(t) \approx C \frac{km}{2\pi} F_R(r) e^{-t/T_2} \sin \left[\omega_0 t + \psi(x) - kr - \theta + \alpha(r) \right]. \tag{42a}$$

Considering the time delay t_d in the signal reaching the receiver, eqn (42a) could be rewritten as

$$i(t) = C \frac{km}{2\pi} F_R(r) \, e^{-\overline{t-t_d}/T_2} \sin\left[\omega_0(t-t_d) + \psi(x) - \theta + \alpha(r)\right] u(t-t_d), \tag{42b}$$

where the time delay is defined as $t_d = r/v$ and $u(t)$ denotes the step function.

5.4.1   Ensemble of signals from radiating regions

A sample under test will have numerous radiating regions, but to show peculiar effects due to the individual radiating regions, the ensemble of signals obtained is modeled as a discrete summation. The signals thus obtained may be expressed as

$$x(t) = C \frac{km}{2\pi} \sum_i F_R(r_i) \, e^{-\overline{t-t_{di}}/T_2} \sin\left[2\pi f_0(t-t_{di}) + \psi(x) - \theta(i) + \alpha(r_i)\right] u(t-t_{di}), \tag{43}$$

where $F_R(r_i)$ is the magnitude modifying factor, which depends on the distance between the radiating region i and the receiver; t_{di} is the time delay in the signal reaching the receiver; $\psi(x)$ is a constant phase; $\theta(i)$ is a phase angle depending on the placement of the receiver coil; $\alpha(r_i)$ is a distance dependent phase term due to the complex nature of the near fields and $u(t)$ is a step function. The integer i will vary over the number of radiating regions.

In MR experiments, signals are obtained using frequency and phase encoding field gradients, in which case the signal described in eqn (43) should be modified to include different frequencies and phase. Thus eqn (43) becomes

$$x(t) = C \frac{km}{2\pi} \sum_i \sum_j F_R(r_i) \, e^{-\overline{t-t_{di}}/T_2}$$
$$\times \sin\left[2\pi f_i(t-t_{di}) + \psi(x) - \theta(i) + \alpha(r_i) + \phi_j\right] u(t-t_{di}), \tag{44}$$

where ϕ_j is the phase angle due to the phase encoding gradient. Thus, the signal of eqn (44) is encoded in frequency and phase. The frequency f_i and phase ϕ_j are defined as

$$f_i = f_0 + g_f i = f_0 + \Delta f, \tag{45}$$

$$\phi_j = g_p j, \tag{46}$$

where g_f, g_p are the frequency and phase encoding gradients respectively and i, j are aliases for distance used here in a discrete sense and span the number of frequency and phase encoding steps.

For illustrative purposes, nine discrete radiating regions are selected on a square, as shown in Fig. 5.3. This setup has been selected to study the effects of the frequency and phase encoding gradients on the near-field MRI signals and also to show the effect due to the magnitude modifying factor in modulating the signal strength from region to region.

The receiving coil is placed perpendicular to the x-axis. This setup simplifies the calculations. The radiating regions are assumed to be resonating at 100 MHz, corresponding to $\gamma = 50 \times 10^6$ at a field strength of 2 T. These signals are stepped down in frequency by coherent quadrature demodulation which enables the phase information to be retrieved. For this purpose, the local oscillator frequency was assumed to be 99.5 MHz. Thus, the center frequency is 0.5 MHz (500 kHz). The relaxation time constant T_2 was assumed to be 0.003 s, a realistic value closely matching a practical situation. The other phase angle $\alpha(r)$ is calculated from eqn (101) in Chapter 2. The magnetization density m is assumed to be 4.4×10^{-3} Am^{-2}. When a field gradient (g_f) of 1 mT/cm is applied, the corresponding change in frequency can be calculated from

$$\Delta f = \frac{\gamma \cdot g_f}{2\pi}. \qquad (47)$$

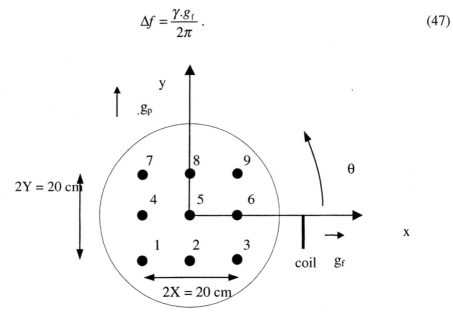

Figure 5.3: Radiating regions under study.

Substituting the values of γ and g_f, we get $\Delta f \approx 8750$ Hz/cm. Similarly, a phase gradient of 3.75°/cm is assumed. Thus for the radiating regions considered, the frequency difference between regions (1, 4, 7), (2, 5, 8) and (3, 6, 9) in Fig. 5.3 will be

$$\Delta f = 8750 \times 10 = 87.5 \text{ kHz}.$$

The relative phase difference between regions (1, 2, 3), (4, 5, 6) and (7, 8, 9) will be
$$\Delta\phi = 3.75 \times 10$$
$$= 37.5°.$$
The signals from the different regions are shown in Fig. 5.4.

In Fig. 5.4, the signals in each column have identical frequency and it is progressively increasing from the left to right column. This is due to a positive frequency-encoding field gradient in the x direction. The starting phases of the signals in a row are approximately identical. This is because the phase-encoding field gradient is applied in the y direction and it does not affect the signal phase in the x direction. The small perturbations are due to the additional phase quantities as defined in eqn(43). This includes the distance dependent *phase modifying factor* too. The resultant signal is superimposed on the signal from region-3 (bottom right corner of Fig. 5.4). It is seen that the amplitude of the signal from region-6 determines the magnitude of the resultant signal. For clarity the resultant signal is shown in Fig. 5.5 separately.

The nine regions were assigned equal radiating strengths, i.e. equal magnetization density. Hence the initial amplitude of the received signal has to be equal. But as seen from Fig. 5.4, it is not. This is due to the effect of the *magnitude modifying factor*, $F_R(r)$. Depending on the distance between the coil and the radiating regions, the amplitude of the signal is changed. As demonstrated in Fig. 2.9, the value of factor $F_R(r)$ is large when the distance r is small. Hence the amplitude of the signal from region-6, which is the closest to the coil, is the largest. The difference in the signal amplitude from different regions could be as large as 14 dB. In the more complete signal model proposed in eqn (41), it is possible to account for the additional phase and amplitude perturbations due to the near-field phenomena.

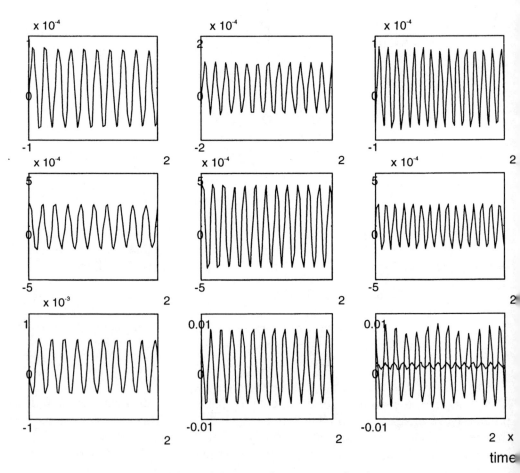

Figure 5.4: Magnetic resonance signals.

5.5 Magnetic resonance image results

In order to see the effects of the additional phase and amplitude modulation factors on image synthesis, the nine radiating regions shown in Fig. 5.3 are imaged using Fourier imaging techniques described in Section 5.4.2. We ignore the effects of the complex impedance of the different materials present in the human body and the interaction of these materials with the resonating magnetic dipoles. The complex impedance of the biomaterials and the net effect due to the interactions can be considered as a random phase noise and compensated. The simplified signal model that ignores this random phase noise is used to illustrate the effects due to the near EM fields in MRI in particular and biomedical imaging in general. In Hoole (2000) we proposed a wavelet based filter to eliminate noise in MR signals.

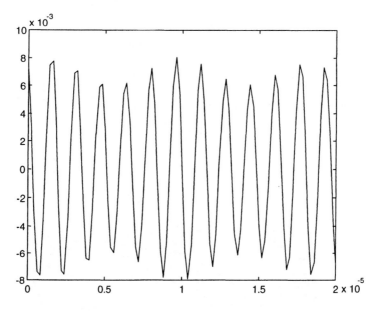

Figure 5.5: Resultant magnetic resonance.

5.5.1 H_r based image

The image of the nine radiating regions, reconstructed using Fourier imaging techniques, when the signal is modified by the MMF of the radial magnetic field component, is shown in Fig. 5.6(a) and (b). Although the intensity of the nine regions should be equal as per the assigned magnetization density values, it is not identical.

The relative image intensity of regions 1–9 from Fig. 5.6 is 1:1.71:4:1.14: 2:5.43:1:1.71:4. The relative error in image intensity is as large as 9 dB between the extreme values. This will render the image unreliable for statistical studies, segmentation and classification.

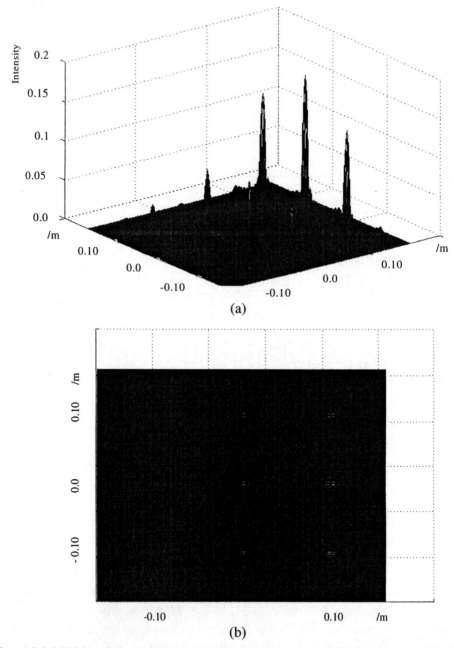

(a)

(b)

Figure 5.6: MRI imaging using H_r field. (a) Perspective view of H_r based image. (b) Plan view of the image.

5.5.2 H_θ based image

Figure 5.7 shows the H_θ based image of the nine regions described in Fig. 5.3. The relative image intensity of regions 1–9 is 1:1.3:2.24:1.05:1.38:2.95:1:1.3:2.24 respectively. The maximum image intensity error is small when compared with that for the H_r based image. The maximum relative error was calculated to be less than 5 dB. Hence it is seen that the images obtained with H_θ based signals are better than that obtained with H_r signals.

5.5.3   Image due to circularly polarized signals

In the near-field region of the electromagnetic signals, two orthogonal magnetic fields H_r and H_θ are present. In MRI technology the receiving coils are usually placed in the near-field region. Hence the resultant received signals are circularly polarized. If the receiver is capable of picking up the circularly polarized signal, which is the current practice, the MR signal magnitude will be modified by the resultant MMF defined by eqn (99) in Chapter 2. The image obtained for the nine-region test case with such a signal model is shown in Fig. 5.8.

The relative image intensity ratio for regions 1–9 is 1:1.7:3.75:1.10:1.95:5: 1:1.7:3.75. The maximum deviation in intensity is 7.2 dB. It is apparent that the response resembles the image due to H_r. In all the three cases considered with H_r, H_θ and $\sqrt{(H_r^2 + H_\theta^2)}$, the images do not represent the information accurately.

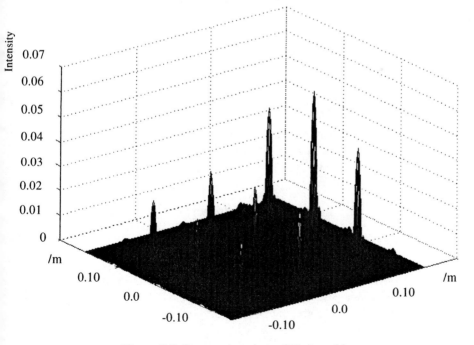

Figure 5.7: Perspective view of H_θ based image.

Based on the simulation and theoretical studies carried out, it is established that the images obtained through imaging routines currently in use do not convey the information accurately. The image misrepresentation can be classified as image intensity inhomogeneity. This error has been unaccounted for because of the incompleteness of the underlying signal model used and its implications to the imaging routine have not been analyzed. Although an image intensity inhomogeneity due to the

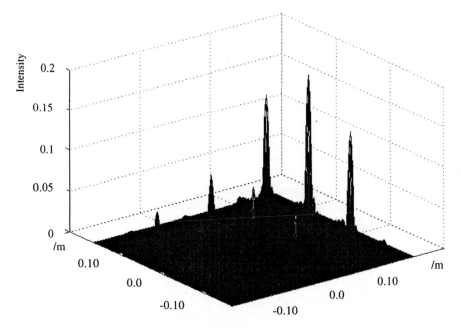

Figure 5.8: Image due to circularly polarized signal.

external magnetic field inhomogeneity has been analyzed, correction measures proposed in the literature have ignored the near-field effects reported in this section.

Since a more complete signal model has been proposed in this thesis, it is possible to account for this additional error in intensity and perform a correction for the image intensity inhomogeneity. This procedure is explained in the next section.

5.6   Correction for intensity inhomogeneity artifact

5.6.1   Correction for phase modifying factor

Erroneous phase information will distort the image geometry and intensity. The radiating regions in an image will be misregistered in position, modulated in intensity and scaled in size. It has been noted that the phase information is corrupted due to inhomogeneity in the B_1 field and due to the susceptibility change near bone–tissue and tissue–air interfaces. A variety of correction procedures have been proposed. One way of correcting images is as follows: a set of reference scans has been obtained with homogenous material as subject and the correct phase information thus derived has been utilized to correct the distorted phase information in the actual routine. Images obtained through this procedure are seen to be satisfactory, but the distance dependent phase information has not been removed by this procedure and will distort the image. Phase can be corrupted through coil imperfection and alterations to its nominal position with respect to the patient. This image distortion has been termed as *coil effect* and attributed to a bias field. Although this effect can be ignored for qualitative analysis, it is important for automated image segmentation activity. The bias field has been estimated and corrected using phantom images and polynomial fitting. But these methods are based on assuming that the corrupting effects are patient independent, which is not correct. It has been found that the correction should be done retrospectively and also that post-filtering is better than pre-filtering. Our approach in correcting for the phase and magnitude modifying factors is post processing steps and thus highly ·suitable for this problem. In addition, image artifacts are created through phase distortion due to static field inhomogeneity, relative motion of the patient and $N/2$ ghost artifact due to non-optimum data acquisition system performance.

The variation of PMF due to H_r and H_θ with distance, as shown in Fig. 2.8, is not rapid. As such, the phase variation between signals from adjacent regions is not marked. Thus there are imperceptible intensity and mispositioning differences in the images obtained. The image domain data in this thesis have not been corrected for phase perturbations, but corrected for the more notable changes in intensity due to the MMF. If correction is to be incorporated, the correction for PMF should be performed before the correction for MMF.

5.6.2   *Correction algorithm for magnitude modifying factor*

The knowledge of the underlying near-field MR signal model enables a correction procedure to be incorporated in the MRI imaging routine. This procedure is to operate on the distorted image and the algorithm is shown schematically in Fig. 5.9.

The procedure could be explained as follows:

- The image is formed using the Fourier imaging technique, as shown in the upper first two blocks in Fig. 5.9. The imaging routine determines the resolution and the field of view. Hence the parameters m, n and X, Y are known, where m, n are the number of points processed in the x, y directions respectively, X, Y are the object dimensions in the x, y directions respectively with reference to Fig. 5.3. These data are fed to the MMF black box.
- The orientation and distance of the coil with respect to the fixed axis system, as shown in Fig. 5.3, are also fed to the MMF black box. The orientation could be detected by the signal pickup electronic system. The distance is usually a fixed parameter. Depending on coil orientation, an appropriate signal model is chosen.
- The MMF is computed and fed to the post processor, which incorporates this factor to correct the distorted image. The end result is the corrected image.

The MMF and post processor blocks are discussed in Sections 5.6.3 and 5.6.4 respectively.

5.6.3   MMF block

The input parameters and the function of the MMF block could be further detailed as shown in Fig. 5.10.

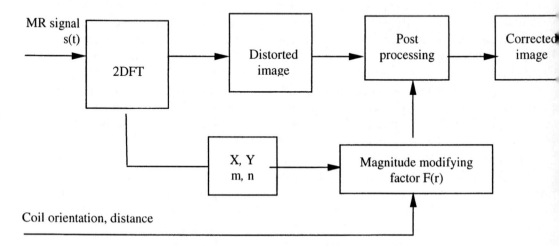

Figure 5.9: Schematic diagram of the correction process.

Depending on the orientation of the pickup coil, the receiver picks up the H_r, H_θ or the circularly polarized resultant signal. When the coil receives the H_r component, there will be contribution from the H_θ component as well. Hence the usual signal to be used in processing is that due to the resultant, and it is evident from image results and analysis of the MMF that the resultant factor follows the response of H_r. Hence the coil orientation should be selected correctly for optimum performance.

The field of view parameter gives the value of X and Y. Further, the number of points used for the 2DFT operation (m, n) is known since it is set by the operator. The calculation is done using

$$x_p = \left(\frac{2X}{m}\right)p, \qquad y_q = \left(\frac{2Y}{n}\right)q, \tag{48}$$

where p, q are the parameters of the image domain data $I(p, q)$. The distance to the radiating region from the receiver is calculated from

$$r_{pq} = \sqrt{\left(d - x_p\right)^2 + y_q^2}, \tag{49}$$

where d is the distance of the coil center from the origin of the axis system. Once the distance is known, the corresponding factor could be calculated for each point on the image. Thus the output of the MMF block is a spatial filter $S(p, q)$.

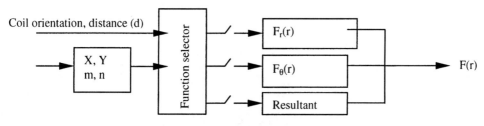

Figure 5.10: MMF block.

To reduce the computational effort, the image could be segmented and the filter coefficients computed for only a fraction of the total number of image pixels. This is feasible since there will not be abrupt intensity changes within a few pixels in theimage. Also the circular symmetry of the problem (if applicable) could be used to halve the number of points required for distance calculation. Hence for a 128×128 order filter, calculations on 64×64 points are sufficient. The filter coefficients could be saved on computer disk and need not be computed every time.

The spatial filter is given by

$$S(p, q) = 1/F(r_{pq}). \tag{50}$$

Such a filter response is shown in Fig. 5.11. Here the field of view and the number of points for resolution calculation were taken to be 20×20 cm and 128×128 respectively.

Equal coefficients are shown with the same color on the graph. The color changes gradually signifying the fact that the coefficients do not change rapidly. Also the filter coefficients are symmetrical about the x-axis. Due to this property of symmetry, the computational requirement could be reduced by half as mentioned earlier.

5.6.4 Post processing

The corrected image domain data $I'(p, q)$ could be obtained by the following operation:

$$I'(p,q) = I(p,q) S(p,q), \quad p, q: -\tfrac{1}{2}N, \ -\tfrac{1}{2}N+1, ..., \tfrac{1}{2}N-1, \tag{51}$$

where the order of the spatial filter is identical to that of the image domain dimension N. When a reduction factor R is used on the order of the filter, the operation could be defined by

$$I'(p,q) = I\left(\frac{p}{R},\frac{q}{R}\right)S(a,b), \quad \text{ceiling, } (p/R, q/R): 1, 2, \ldots, N/R, \quad a, b: 1, 2, \ldots, N/R$$

(52)

where ceiling gives the nearest upper integer value. The resolution afforded by the corrected image depends on the order of the spatial filter.

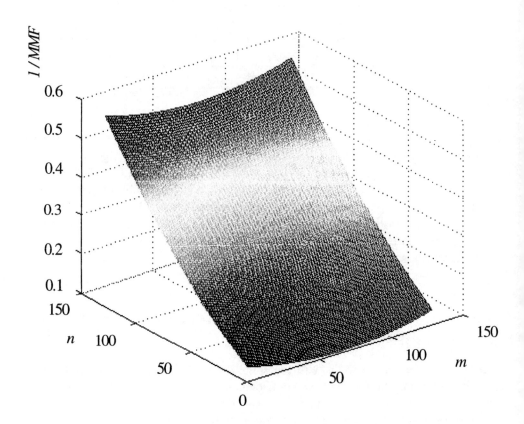

Figure 5.11: Spatial filter of order 128×128 for a field of view of 20×20 cm.

5.6.5   Corrected image

The distorted image of Fig. 5.8 was filtered by the MMF filter and the resultant image is shown in Fig. 5.12. The nine radiating regions shown in Fig. 5.12 have equal radiating strengths after the filtering process. This image is more useful for clinical diagnosis and statistical studies since the image of Fig. 5.12 does not have the intensity inhomogeneity artifact. Using the near-field signal modeling reported in Section 5.4, it has been possible to overcome an inherent intensity inhomogeneity problem associated with the MR images.

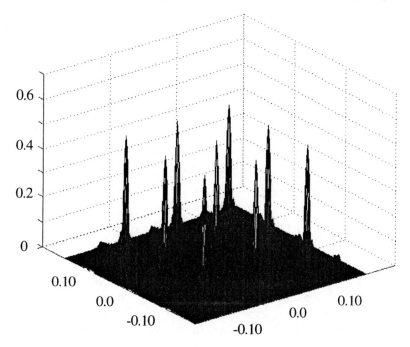

Figure 5.12: Corrected MR image of nine radiating regions.

5.7   Coil antennas used in MRI systems

5.7.1   Receiver coils

An alternating magnetic field obtained in an MRI scenario should be converted to a voltage to be operated upon. These alternating magnetic fields should be created to excite the samples. A coil performs this operation. Although an antenna does the same function, it acts in the far fields whereas a coil operates in the near fields. This section is based on Chen and Hoult (1989), from which the figures are also adapted with permission.

5.7.1.1   Coil design
A coil used in MRI should be designed to produce the B_1 field in the xy-plane. Design analysis based on mutual coupling, skin effect and frequency of operation has established that to achieve optimum operation the number of turns in a coil should be less than one. Hence radio frequency antenna design principles are modified to suit the design constraints of these inductor coils.

A Hertzian loop is shown in Fig. 5.13. The magnetic field is inhomogenous and the electric field strength is high within the coil. The near fields of the coil, rather than the far (radiation) fields, are used in MRI. The high electric field strength should be eliminated as the lossy dielectric tissues are heated by the electric field and might cause burns.

Step I: Hertzian loop

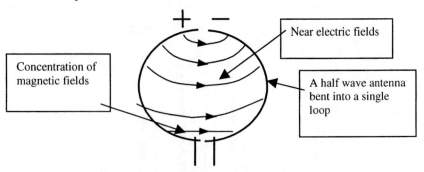

Figure 5.13: Hertzian loop (Chen and Hoult, 1989).

To reduce the dielectric losses a balanced length of wire or a high Q capacitor could be added as shown in Figs 5.13 and 5.14 respectively. Then most of the electric field is confined to the wire or the capacitor and hence dielectric losses due to the electric field are reduced. But this reduces the resonant frequency of the structure.

Step II: High Q capacitor

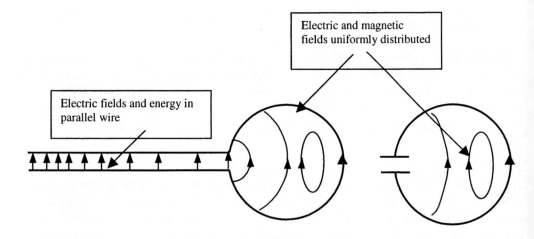

Figure 5.14: Loop with balanced line (Chen and Hoult, 1989) and capacitor (Chen and Hoult, 1989).

Step III: Capacitor distribution

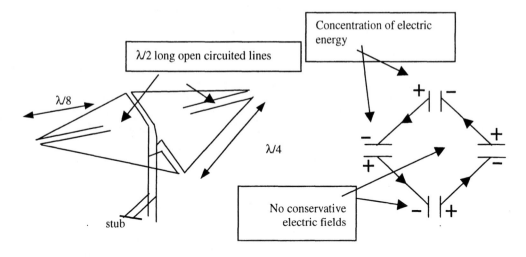

Figure 5.15: Alford loop and equivalent circuit (Chen and Hoult, 1989).

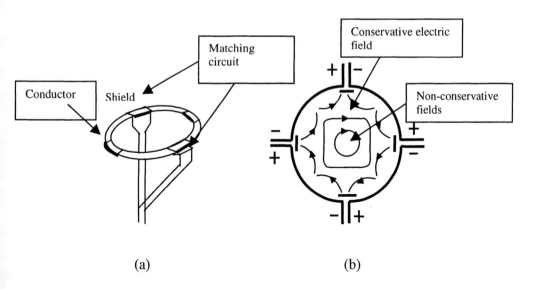

(a)   (b)

Figure 5.16: Surface coil with shields and equivalent circuit (Chen and Hoult, 1989).

The configuration shown in Fig. 5.15 effectively distributes capacitance discretely about the aerial. Thus the resonant frequency increases. The conservative electric field is so redistributed that its presence in the middle of the loop and outside the loop is negligibly small. Effectively, the interference due to near electric field is reduced.

A surface coil with additional shields to further reduce the dielectric losses in the sample is shown in Fig. 5.16(a). The equivalent circuit and associated fields are shown in the adjoining figure.

5.7.2 *Transmitter coils*

An arrangement to produce the rotating magnetic field B_1 on the *xy*-plane is shown in Fig. 5.17. Such an idea is quite familiar to those involved in induction motor design, where two or three phase coils are used to produce rotating magnetic fields.

The current in coils A and B is 90° out of phase. This effectively creates a magnetic field that rotates at the coil center at the frequency of the current. This arrangement can be modified to be used as a receiver as well. A coil used in imaging the head is shown in Fig. 5.18.

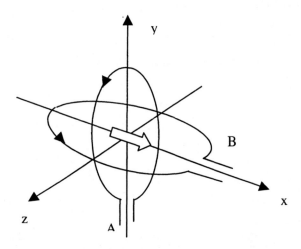

Figure 5.17: Quadrature coils with current 90° out of phase (Chen and Hoult, 1989).

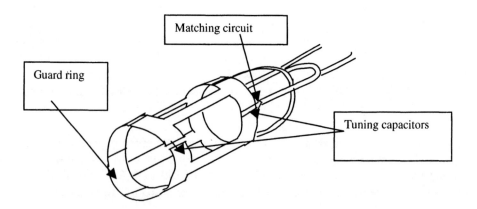

Figure 5.18: Alderman and Grant head coil (Chen and Hoult, 1989).

6 Synthetic aperture antennas and imaging

Tan Pek Hua, Dennis Goh, P.R.P. Hoole and U.R. Abeyratne

6.1 Basic principles of radar signal processing

6.1.1 Introduction

Moving a single element antenna over space generates a synthetic aperture antenna. An example of this is an aircraft-mounted antenna that operates (i.e. transmits and receives signals) as the aircraft moves at a velocity of 300 m/s, say. In such a case, over a time period of 1 s the antenna will be at 30 discrete points of 10 m apart. We now have an artificial array antenna of 30 elements with an interelement distance of 10 m. This is called a synthetic array or synthetic aperture antenna. We get such a situation in mobile communication systems too, where the mobile phone moves in space as it communicates with the base station. Hence the mobile phone antenna is a synthetic aperture antenna. In this chapter, we shall illustrate the principles of synthetic aperture antennas by looking at its performance in a radar system. A radar system is a system in which an antenna transmits a pulsed signal, then pauses before transmitting another pulse to collect the reflected return pulse of the first signal. The return pulse signals are processed to obtain details like distance and velocity of the target, as well as an image of the target. The synthetic aperture antenna allows us to perform such powerful processing to do remote sensing.

A radar system is made up of a transmitter (to generate the pulse to be radiated out), a receiver (to pick up the return signal), an antenna which is physically moved (in addition its beam may be electronically steered in some cases) and a signal processor to get one-dimensional and two-dimensional (image) details of the target that was observed. The transmitter generates a sequence of pulses (or bursts) which is launched out towards the target by a narrow beam antenna. The pulsed signals form electromagnetic waves in space and travel at the velocity of light ($c = 3 \times 10^8$ m/s) in free space. In media other than free space with a relative permittivity of ε_r (e.g. 80 for seawater), the velocity of the waves will be reduced to $u = c/(\varepsilon_r)^{1/2}$. The waves hit the target, and part of the wave energy is reflected back towards the radar antenna. These reflected waves are the return signals picked up by the radar antenna and processed by the receiver. The time delay, τ, between the transmitted and return pulse is a measure of the two-way distance between the radar antenna and the target. This two-way distance is twice the range of the target, where range is the line-of-sight (LOS) distance between the radar antenna and the target given by

$$\text{Range} = \frac{u\tau}{2}, \tag{1}$$

where u is the speed of the electromagnetic wave and τ is the time delay or echo of th return signal.

Since the beam of the antenna is narrow and pointed in a known direction, th amount of wave energy reflected back from the target will depend on the direction ar the reflectivity (reflection coefficient) of the target in that direction. This variation ma be used to determine the direction in which the target is situated. Now using the da on the range and angular direction of the target, we may reconstruct or synthesize th position and reflectivity map of the target. The issue becomes more complicated whe there are two or more targets falling into the antenna beam at the same time. In such case the reflected signal carries energy of waves reflected by both objects, an depending on the distance between the two targets, the time difference between the tw reflected signals may not be sufficiently large to separate the two targets. This rais the issue of ambiguity, since it is not clear which of the two targets or reflecting poin is represented by the return signals. Thus a central issue becomes spatial resolution th could be achieved from the return signals; the return signals should be processed achieve minute range and azimuth resolutions.

It can be shown that range resolution is inversely proportional to the bandwidth the transmitted signal. Hence a signal with a wide bandwidth will help to achieve fin range resolution. One way to achieve resolution in the azimuth direction is to use a antenna with a narrow azimuth beam width. Narrow antenna beam width means larg antenna aperture size or large radar frequency. One practical constraint on lar; antenna length is the difficulty of mounting a physically large antenna on an aircra Increasing transmitted frequency means greater adverse effects of the mediums: frequency is increased above say 20 GHz, the transmitted and return signal will significantly attenuated by air or any other medium through which the signals trave Hence to improve azimuth resolution, we have to resort to synthetic array or aperture.

6.1.2   Synthetic aperture radar

In synthetic aperture radar (SAR) a moving radar collects many pulses of retu signals, and by comparing them achieves better azimuth resolution. The azimu resolution obtained is much finer than that of a stationary antenna with the same re beam width; the moving of the antenna artificially extends the length of the anten and hence narrows the overall azimuth beam width. By illuminating the target different angles and different locations, an image (two-dimensional or thre dimensional) of the target could be obtained.

Figure 6.1 illustrates the three common SAR imaging modes, namely the spotligh stripmap and scan modes.

6.2   Inverse synthetic aperture radar

The inverse SAR (ISAR) mode is a fourth mode of synthetic aperture antenna imagir here the antenna is kept stationary at one point and the object is moved around t antenna. As the object moves around the antenna, different views of it are presented the antenna, so that the return pulses carry back a complete range of information on the entire target. The principles governing SAR and ISAR are the same since what is basic is the relative motion between the antenna and the target. In the case of ISAR the moving target re-radiating back return signals acts as a synthetic array antenna. In this chapter, the ISAR scenario is described. The case study considered is that of imaging an aircraft in flight by a stationary antenna.

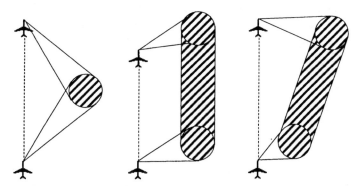

Figure 6.1: Three common SAR imaging modes: spotlight, stripmap and scan, respectively. (Reprinted with permision from Spotlight Synthetic Aperture Radar Signal Processing Algorithms, by W.G. Carrara, R.S. Goodman and A. Majewski's. Artech House, Inc., Norwood, MA, USA. www.artechhouse.com (Carrara et al., 1995).)

6.3   One-dimensional imaging with point scattering

6.3.1   Overview

This section is intended to give the reader an overview of the one-dimensional imaging process before going on to the mathematical description. Consider M point targets or M number of scattering points on a single target, as shown in Fig. 6.2. We assume that they all lie in a straight line. The distances separating two adjacent points are unequal.

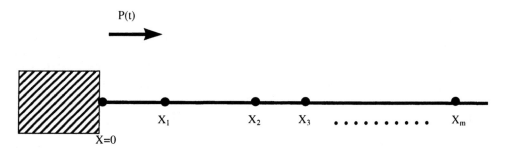

Transmitter/Receiver

Figure 6.2: Geometry for one-dimensional imaging.

An antenna transmitter located at $x = 0$ radiates a pulsed signal $s(t)$ of finite duration T. The leading edge of the $s(t)$ reaches the first scattering point at time $t_1 = x_1/u$, where u is the speed of wave propagation in the medium where the targets reside. The first scattering point reflects back a   portion of the signal $s(t)$ back to the antenna receiver reaching $x = 0$ in t_1 seconds later. Assuming that the reflected signal is very small in comparison to the strength of the transmitted signal $s(t)$, the same signal strength $s(t)$ now hits scattering point 2, and later 3 and so on until it hits point m. Each scattering point will reflect back a portion of $s(t)$ depending on the reflection coefficients of the points. Figure 6.3 shows the transmitted pulse and the m number of reflected pulses. We have ignored the attenuation of the signal due to the transmission medium (e.g. air) and beam spreading.

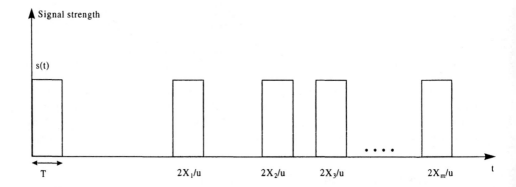

Figure 6.3: Transmitted and returned pulses.

If the scattering points are close together, the returned pulses may overlap each other as shown in Fig. 6.4, in which case it will be difficult to resolve between the scattering points at x_2 and x_3.

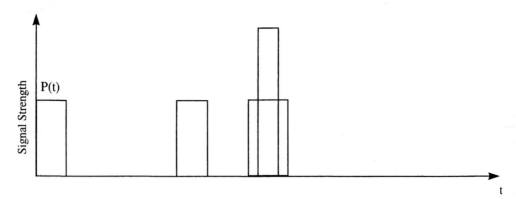

Figure 6.4: Overlapping returned pulses.

It is seen from Fig. 6.4 that if we shorten the pulse duration T, then we can get better resolution by avoiding overlapping return pulses. Of course shortening the pulse has the disadvantage of reducing the energy in the signal and hence the distance up to which we can still detect the return signal. The requirement is that closely spaced points can be resolved if they are separated in time delay by at least the width of the transmitted pulse, T. If the distance of separation between two adjacent scatterers is less than $uT/2$, then the return pulses will overlap, making it impossible to determine where one point ends and the other begins. The distance $uT/2$ is called the range resolution of the radar. As noted, reducing T cannot be done without the overall consideration of the maximum distance to which we require the radar to perform. The transmitted radar pulse should carry more energy if the target is very far, so that the little energy reflected back can be processed by the receiver electronics without getting swamped out by the electronic noise. Hence an alternative is to design a pulse shape that has sufficiently short time duration while having the required energy, and may be processed to distinguish different scatterers. The pulse should be designed such that

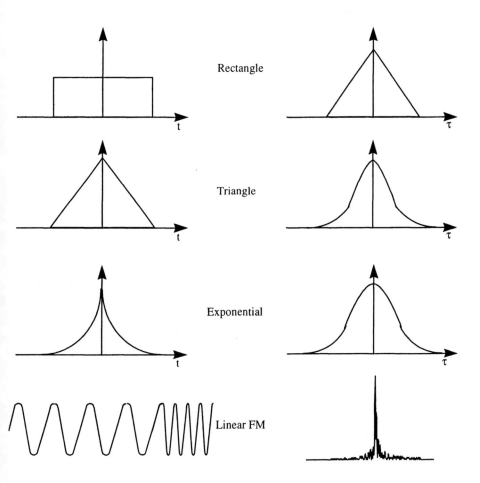

Figure 6.5: Autocorrelation of several functions (from Fitch, 1988).

overlapping returns from different scatterers can be separated. This means that the correlation between the pulses should be small everywhere except when the time delay between signals is zero. From the autocorrelation of several functions shown in Fig 6.5, we note that one of the best functions to use as waveforms inside each pulse is the linear FM pulse, also known as the chirp pulse.

Since the linear FM or chirp pulse can be compressed into a sharp pulse, even when there is an overall overlap between two return pulses, processing the return pulses by a matched filter will enable us to separate two overlapping returns as they each become compressed into sharp spikes on autocorrelation. Thus, the range resolution can be further enhanced by the compression of the chirp pulse through a matched filter. Then each transmitted pulse is a pulse of linear FM signal, which is a kind of a wavelet. The reader would now realize that designing an antenna system to perform complex activity like imaging requires not only a synthetic array or synthetic aperture antenna, but also a waveform that is well shaped and suited for the particular application. In telecommunication systems, we use other kinds of signal shaping or modulation to improve on the system performance. Hence antenna design cannot be considered independently of the best suited signal for complex systems.

Linear FM (chirp) waveform is used as the transmitted radar pulse. As mentioned in the previous section, linear FM is used so as to reduce the superposition effect of the two return pulses or echoes which differs by a small time delay. The linear FM waveform is characterized by the equation

$$S'(t) = 2\pi(ft + 0.5at^2), \tag{2}$$

where f = initial frequency of the linear FM and a = chirp rate or the rate of change of frequency.

This basic signal is up-converted to a higher frequency known as the carrier frequency f_c by mixing, before transmission. The higher frequency is required to improve transmission and reception of the wave. Hence, the final transmitted wave is given by

$$s(t) = \cos[2\pi(f_c t + 0.5at^2)] \; \text{rect}((t-T/2)/T), \tag{3}$$

and it is shown in Fig. 6.6.

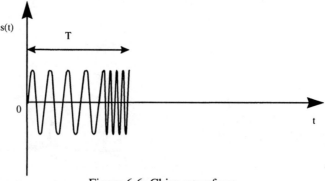

Figure 6.6: Chirp waveform.

The definition of the rect( ) function in eqn (3) is as follows:

$$\text{rect}((t-T/2)/T) = \begin{cases} 1, & 0 \le t \le T, \\ 0, & \text{otherwise.} \end{cases}$$

Now consider the reflection of the transmitted signal off an object at distance R from the antenna. The received return pulse is the result of $s(t)$ and its reflected portion having traveled a distance $2R$ from transmission to reception. This two-way distance corresponds to a time delay of $\tau = 2R/u$ and the received signal is of the form $\sigma \cdot s(t-\tau)$, where the factor σ (reflection coefficient) is a function of the target geometry and material property at carrier frequency f_c.

By mixing (down-conversion) the return signal picked up by the antenna, the down-converted signal r_{if} is given by

$$r_{if}(t) = \sigma \cdot \cos[2\pi(-f_c\tau + f_{if}t + 0.5a(t-\tau)^2)] \, \text{rect}((t-T/2-\tau)/T). \qquad (4)$$

The return signal $r_{if}(t)$ is processed by a matched filter, the compressed output will be scaled by some σ and located at a time delay of τ as shown in Fig. 6.7.

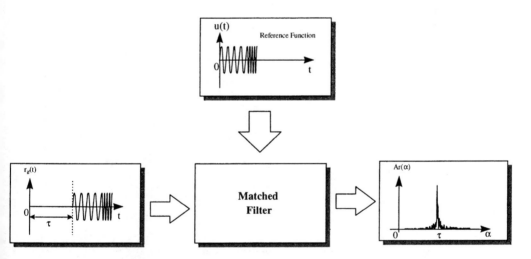

Figure 6.7: Range compression by matched filter.

The matched filter correlates two signals. It correlates the return signal with a reference signal. The reference signal used here is a replica of the transmitted linear FM signal $s(t)$. Based on the mixed down signal of eqn (3), the appropriate reference pulse for the matched filter is therefore given by

$$s_{ref}(t) = \cos[2\pi(f_{if}t + 0.5at^2)] \, \text{rect}((t-T/2)/T). \qquad (5)$$

Thus, the output $A(\alpha)$ of the matched filter is the correlation of two signals:

$$A(\alpha) = \int r_{if}^*(t) s_{ref}(t+\alpha)\, dt$$

$$= \int e^{-j2\pi(-f_c\tau+f_{if}t+0.5a(t-\tau)^2)}\, e^{j2\pi(f_{if}(t+\alpha)+0.5a(t+\alpha)^2)}\, dt. \qquad (6)$$

Consider $\tau = 0$ first to get the mathematical equation of the output of the matched filter.

$$A(\alpha) = e^{j2\pi(f\alpha+0.5a\alpha^2)} \int e^{j2\pi a\alpha t}\, dt$$

$$= e^{j2\pi a\alpha[f+a(\alpha+0.5t)]} \frac{\sin\left[\pi a\alpha\left(T-|\alpha|\right)\right]}{\pi a\alpha} \qquad \text{for } T \leq \alpha \leq T. \qquad (7)$$

The general shape of this function is easier to see when eqn (7) is rewritten as

$$A(\alpha) = U\,(T-|\alpha|)\frac{\sin[\pi a\alpha(T-|\alpha|)]}{\pi a\alpha(T-|\alpha|)}, \qquad (8)$$

where U is the unit magnitude phase contribution $e^{j2\pi a\alpha[f+a(\alpha+0.5T)]}$. The $U(T-|\alpha|)$ term is a triangle function weighting the $\sin(x)$ over x or the sinc function which follows. The match filtering process is shown in Fig. 6.8.

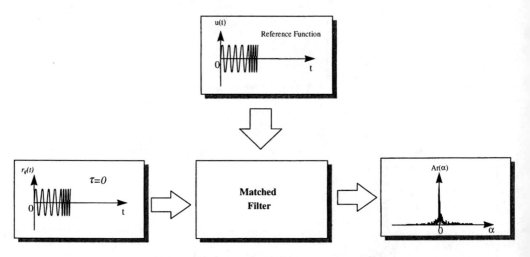

Figure 6.8: Autocorrelation of chirp pulse.

Similarly, if the returned signal is delayed by τ, then by imposing the time shifting property in correlation, the output of the matched filter will be a shifted version of $A(\alpha)$; the shift is by a factor of τ as shown in Fig. 6.9.

6.3.2   Range resolution

The stationary (conventional) SAR or ISAR radar systems all resolve targets in the range dimension in the same way. It is the way in which azimuth resolution is achieved that they differ. Range resolution is achieved through the use of pulsing and time delay sorting as illustrated in Fig. 6.3.

In Fig. 6.3, if T is the pulse width and u is the speed of the pulsed electromagnetic wave, then the range resolution before compression is given by

$$\Delta R = \frac{uT}{2}. \tag{9}$$

This is derived from the fact that to ensure non-overlapping reflections, targets must be separated in time delay by at least the width T of the transmitted pulse. In a conventional pulsed radar, the generation of a pulse of duration T requires a transmitter bandwidth of the order of

$$B \approx 1/T. \tag{10}$$

Hence expressing range resolution in terms of bandwidth we have

$$\Delta R = \frac{u}{2B}. \tag{11}$$

Hence a wide-band radar transmitter and receiver is important to achieve good range resolution. For a linear FM (chirp) pulse of duration T, the bandwidth B is given by

$$B = aT, \tag{12}$$

where a is the chirp rate.

Hence the range resolution after compression is given by

$$\Delta R = \frac{u}{2aT}$$
$$= \frac{uT}{2aT^2}. \tag{13}$$

Comparing eqns (11) and (13), it is seen that range resolution after chirp compression improves by a factor of aT^2. In other words, range resolution is further enhanced by the compression of the chirp pulse through the matched filter, which is what we expect. In other words, a chirp pulse gives a range resolution that is improved by a factor of aT^2.

This can be illustrated by a simple simulation: For a rectangular pulse of duration $T = 0.5$ μs, the range resolution is 75 m as given by eqn (11). But if chirp pulse of the same pulse duration T and chirp rate $a = 3 \times 10^{14}$ is used, this resolution can be improved by 75 times. Thus, the new theoretical range resolution is 1 m (refer to eqn (13)). Figure 6.10 shows two targets separated by 1 m barely resolved. When the targets are 2 and 7 m apart, they can be distinguished clearly as shown in Figs 6.11 and 6.12 respectively. This suggests that the range resolution is about 1–2 m, which is quite close to the theoretical resolution (1 m).

6.3.3   Effect of pulse width variation

Figure 6.13 shows the comparison between the compressed amplitude response to the same point target with different pulse width T.

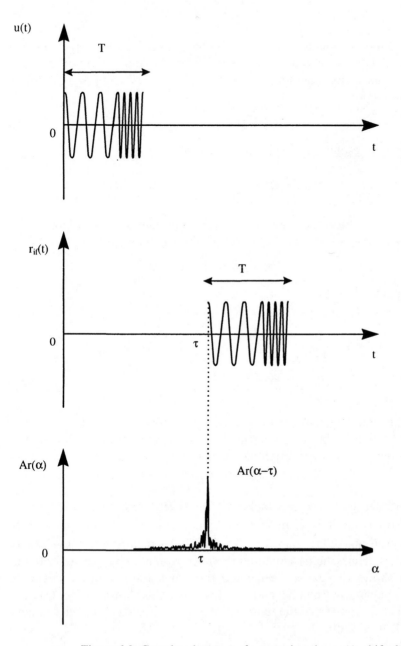

Figure 6.9: Correlated output of a target's echo, $r_{if}(t)$, shifted by τ.

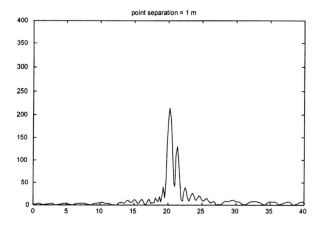

Figure 6.10: Two targets indistinguishable (1 m apart).

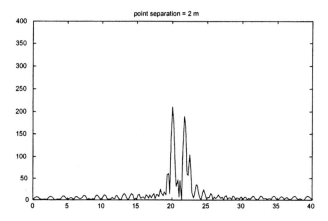

Figure 6.11: Two targets resolved (2 m apart).

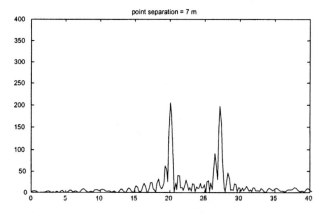

Figure 6.12: Two targets separated by 7 m.

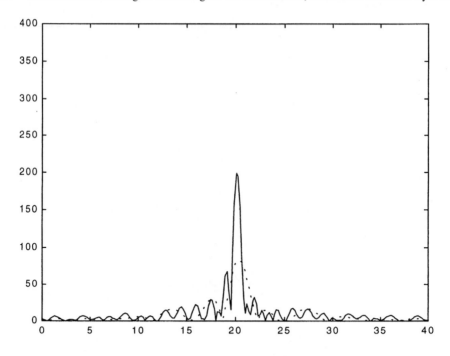

Figure 6.13: Two return pulses with different pulse width T.

The dotted line is simulated with $T = 0.2$ μs while the solid line is $T = 0.5$ μs. The result shows that the main lobe is narrower as the pulse width T becomes larger. It implies that the resolution is finer. This is true because by increasing the pulse duration T in chirp signaling, the bandwidth of the transmitted pulse also increases as shown in eqn (12). As discussed previously, the range resolution increases as the bandwidth of the system increases (see Fig. 6.14). It should also be noted that increasing pulse duration increases the overall transmitted power or, equivalently, improves the output signal-to-noise power ratio. This accounts for the difference in the amplitude of the pulse as shown in Fig. 6.13. This is not true in rectangular pulse signaling, where increasing T improves signal-to-noise power ratio but results in a poorer range resolution.

6.3.4   Effect of a chirp rate variation

In eqn (13), it is evident that the range resolution after compression is dependent on the chirp rate, a. The reason is because by increasing the chirp rate, the bandwidth of the signal also increases (refer to eqn (12)). Since range resolution is inversely proportional to the bandwidth of the transmitted signal, therefore with increasing bandwidth the range resolution improves as shown in Fig. 6.15. It can be seen that the main lobe of the impulse is narrower as the chirp rate a is increased, and hence resulted in a better resolution.

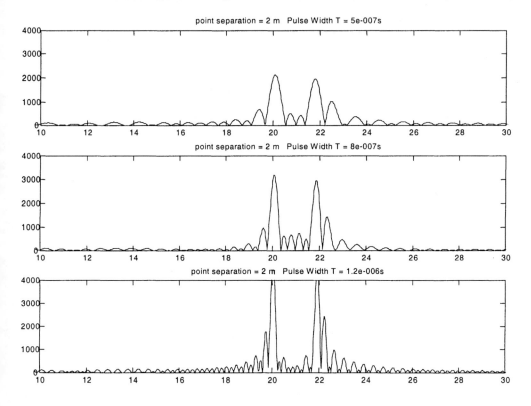

point separation = 2 m   Pulse Width T = 5e-007s

point separation = 2 m   Pulse Width T = 8e-007s

point separation = 2 m   Pulse Width T = 1.2e-006s

Figure 6.14: Resolution improves when pulse duration increases.

6.3.5   Effect of sampling frequency variation

When faced with a lack of resolution in the image, it is natural to attempt to increase the size of the frequency-sampled array by sampling it at a faster rate. As shown in Fig. 6.16, however, this is fruitless: although there are more sampling points, they still cover the same bandwidth. As such, and according to eqn (12), the range resolution remains unchanged. Thus, the addition of more sampling points over the existing bandwidth does not improve resolution.

6.4   Two-dimensional imaging with point scattering

In the one-dimensional imaging discussed earlier, it is assumed that the scattering points are all arranged in a straight line. In general, some points could be displaced in the azimuth direction.

For instance, in Fig. 6.17, point 1 and point 3 have the same relative azimuth position but point 2 is displaced a distance of $A_{12, 32}$ to the right of point 1 and point 3.

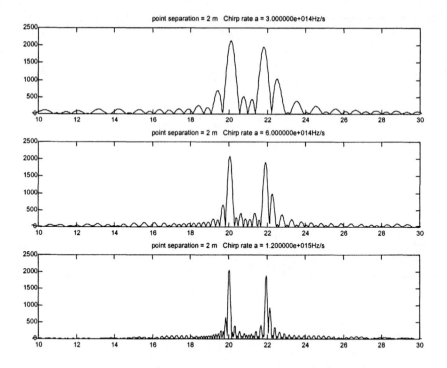

Figure 6.15: Resolution improves when chirp rate increases.

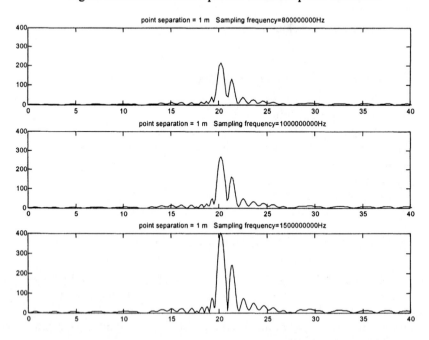

Figure 6.16: Resolution unchanged when sampling frequency increases.

In this case, range imaging alone does not bring out the complete picture (see Range View in Fig. 6.17). It only presents a one-sided restricted view of the positions of the point scatterers and it may be mistaken for three colinear points. Therefore, a two-dimensional picture of the target will give more information in terms of the relative range and azimuth positions of the point scatterers on the target, and the characteristic of the target can be seen clearly.

6.4.1 Overview

The creation of the two-dimensional image is extended from one-dimensional imaging. In order to extract the azimuth information, the target is required to move around the radar. In the imaging process, the target is sequentially illuminated at different positions by the radar beam as the latter travels with uniform velocity in a straight line as shown in Fig. 6.18.

· For each position, the one-dimensional range image of the target (after compression by the matched filter) can be easily obtained. This is repeated at every time interval of $T_{p\text{-}p}$, where $T_{p\text{-}p}$ is the time interval between two consecutive transmitted pulses. For

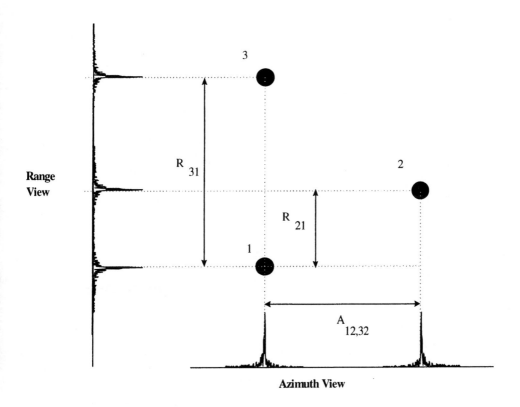

Figure 6.17: Plan view of three point scatterers.

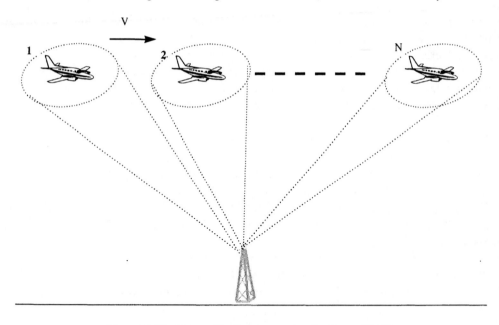

Figure 6.18: Data collection geometry for linear ISAR.

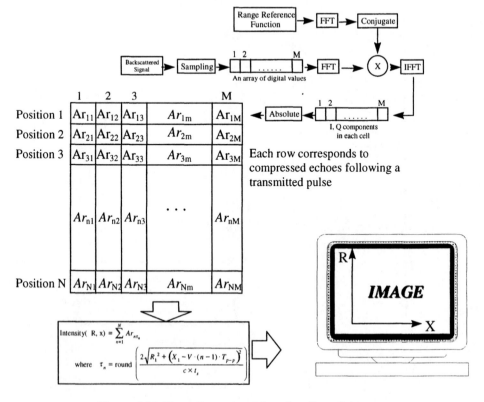

Figure 6.19: Two-dimensional imaging flow diagram.

different positions n, the one-dimensional range image of the target (which is actually an array of absolute numbers Ar_{n1}, Ar_{n2}, Ar_{n3}, ..., Ar_{nm}) is collected and stored into a two-dimensional array as shown in Fig. 6.19.

By a proper processing of this collection of range echoes, the final two-dimensional image of the target can be generated. The detailed procedures of plotting the image of the target is presented in the next section.

6.4.2 Procedures for two-dimensional imaging

6.4.2.1 Data collection

Consider a target moving with uniform velocity V horizontally over the antenna with an aperture L as shown in Fig. 6.20. Assume that there are three point scatterers (a, b and c) on the target and they are arranged in a triangular formation.

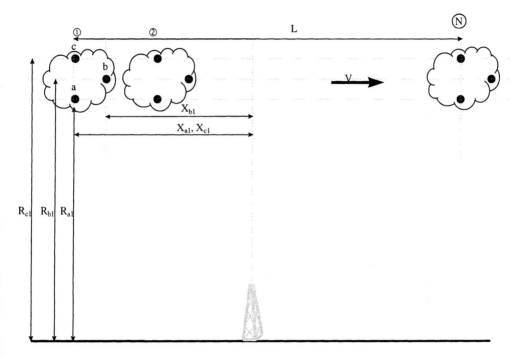

Figure 6.20: Data collection geometry for three scattering points.

When the target is first at position 1, it is illuminated by the radar beam and the return chirp pulse from the three scatterers is compressed into three narrow pulses by matched filtering as shown in Fig. 6.21 (top figure).

The time delays for each point scatterer at position 1 are given below:

$$\tau_{a1} = \frac{2\sqrt{R_{a1}^2 + X_{a1}^2}}{c},$$

$$\tau_{b1} = \frac{2\sqrt{R_{b1}^2 + X_{b1}^2}}{c},$$

$$\tau_{c1} = \frac{2\sqrt{R_{c1}^2 + X_{c1}^2}}{c}.$$

As the antenna would be emitting chirp pulses at regular intervals T_{p-p}, during this time interval, the target has actually traveled a distance of VT_{p-p} to reach position 2. Similarly, another one-dimensional range image of the target at the new position can be obtained (see Fig. 6.21, middle figure). Note that the output of the matched filter is actually a $1 \times M$ array of complex numbers which are then converted to absolute numbers and plotted to give the one-dimensional range image. This process is carried out for N positions. The array of real values for all the N positions is collected and stored into a table as shown in Fig. 6.19. This $N \times M$ array can be plotted and shown in

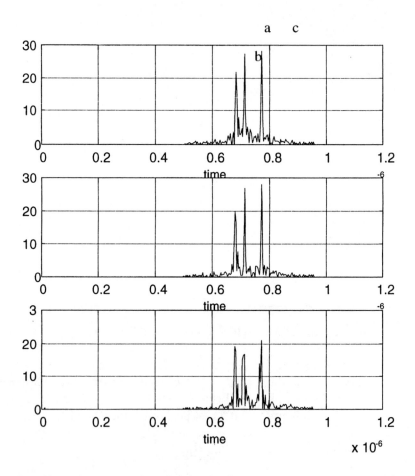

Figure 6.21: One-dimensional range imaging for the first three positions (generated by program Fig.6_21.m).

Fig. 6.22. The plot shows the intensity variations accumulated around three arcs corresponding to each of the three scatterers.

6.4.2.2 Concept for two-dimensional imaging

For a target moving in a straight path with constant velocity V, the general relationship for the time delay of each point scatterer on the target at position n is

$$\tau_n = \frac{2\sqrt{R_1^2 + \left(X_1 - V(n-1)T_{p-p}\right)^2}}{c} \quad \text{for } n = 1, 2, 3, ..., N. \tag{14}$$

This expression has two arguments R_1 and X_1 which represent range and azimuth length at position 1 respectively (see Fig. 6.20). Referring to Fig. 6.20, if the point scatterer "a" located at (40, 100) (refer to Table 6.1) is allowed to move from position 1 to position N, then the corresponding time delay at each position can be calculated using eqn (14). This is shown in Fig. 6.23, where the horizontal axis is the time delay and the vertical axis represents the position number. The simulation parameters provide a useful illustration only; they do not represent any specific radar system.

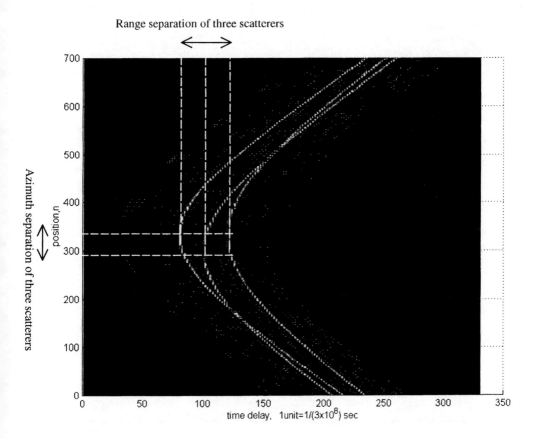

Figure 6.22: Range migration of three scattering points (generated from fig5_22.m).

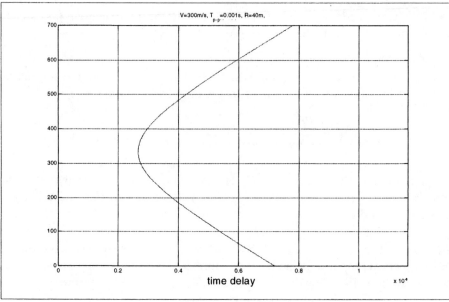

Figure 6.23: Plot of time delay of one point-scatterer.

Table 6.1: Parameters used in the simulation.

Parameters		Descriptions	Values
System			
	f_c	Carrier frequency	1275 MHz
	f_{if}	Intermediate frequency	2 MHz
	T	Chirp pulse duration	0.18 s
	A	Chirp rate	0.8×10^{15} Hz/s
	$T_{p\text{-}p}$	Pulse to pulse interval	1 ms
	N	Number of positions	700
	t_s	Sampling interval	$1/(3 \times 10^8)$ s
	M	Number of sampling points	330
Target			
	V	Target velocity	300 m/s
Three point-scatterers Considered			
Point "a"	R_{a1}	Range	40 m
	X_{a1}	Azimuth	100 m
Point "b"	R_{b1}	Range	50 m
	X_{b1}	Azimuth	90 m
Point "c"	R_{c1}	Range	60 m
	X_{c1}	Azimuth	100 m

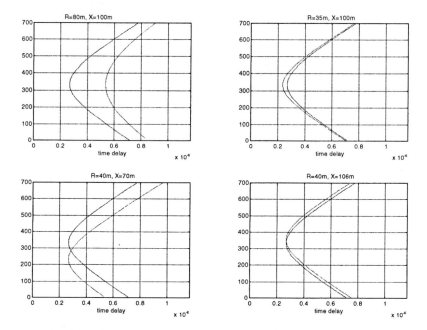

Figure 6.24: Different combinations of R_1 and X_1 values attempting to locate correct values.

Note that the location ($R_{a1} = 40$, $X_{a1} = 100$) of the point scatterer is not known. The task now is to substitute all possible combinations of (R_1, X_1) into the time delay expression in eqn (14) to match the same range migration curve in Fig. 6.23, which is traced out uniquely by scatterer "a". This is shown in Fig. 6.24, where four combinations of R_1 and X_1 values are tried and the corresponding traces are obtained. Hence, this shows that for a particular point scatterer located at (R_1, X_1), the time delay curve traced out is unique and it will only match when the correct value of (R_1, X_1) is used. With this, the location of the point scatterer can be obtained.

6.4.2.3 Development and implementation
This section will show how to apply and implement the concept described above to obtain a two-dimensional image (with range and azimuth) with the $N \times M$ array of data gathered during the data collection stage.

Figure 6.25 shows the geometry and the parameters that will be used in the discussion. In order to produce a two-dimensional image plot of the target, the following information must be provided to customize the plot:

1. Az and Ra – the values of Az and Ra define the size of the plot and they are chosen to cover the target completely. The approximate size of the target of interest is usually known.
2. Δr and Δa – range and azimuth resolutions respectively.
3. A_c and R_c – they define the centroid of the plot. A_c and R_c are given by the position of the centroid of the target where it is first illuminated by the beam (see Fig. 6.25).

Therefore, the plot will contain $Q \times P$ pixels, where

$$P = \frac{Az}{\Delta a}, \qquad Q = \frac{Ra}{\Delta r}.$$

Suppose the parameters shown in Fig. 6.25 assumed the values given in Table 6.2.

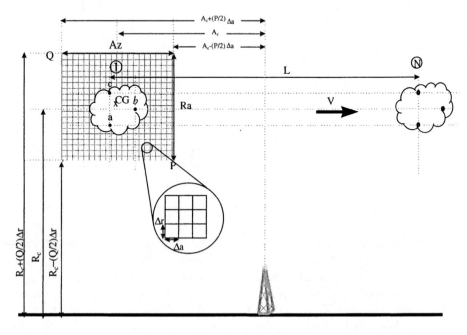

Figure 6.25: Parameters used to customize the plot.

Table 6.2: Values of the parameters used in Fig. 6.25.

Parameters	Description	Values
R_c	Range scene center	50 m
A_c	Azimuth scene center	95 m
Δr	Range resolution	0.5 m
Δa	Azimuth resolution	0.5 m
Ra	Range scene size	50 m
Az	Azimuth scene size	50 m
Q	Number of pixels in range direction	100
P	Number of pixels in azimuth direction	100
V	Velocity of target	300 m/s
T_{p-p}	Pulse to pulse interval	1 ms
L	Synthetic aperture length ($L = VNT_{p-p}$)	210 m

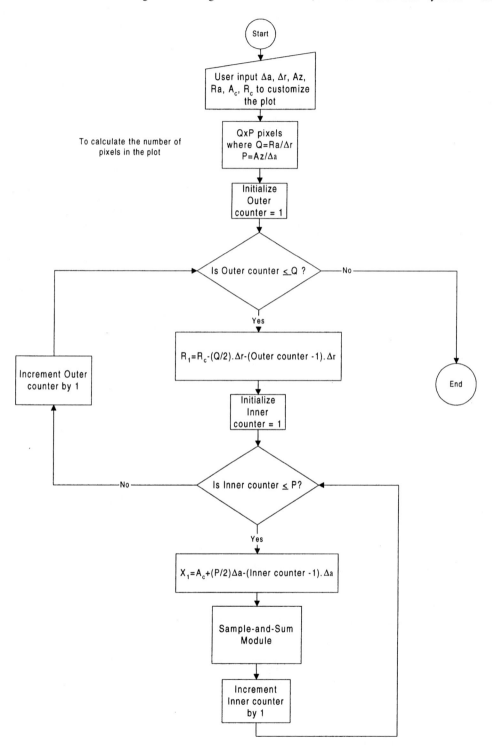

Figure 6.26: Flow chart showing the sequential testing of pixels.

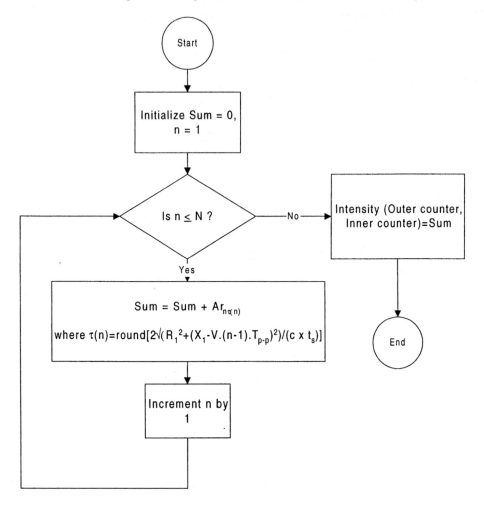

Figure 6.27: Flow chart for range compression.

Recall that during the data collection stage, a matrix of data is obtained as shown in Fig. 6.19. With these data and applying the curve matching concept discussed earlier, the level of intensity at each pixel can be given by

$$\text{Intensity}(R, x) = \sum_{n=1}^{N} \text{Ar}_{n\tau_n} ,$$

$$\text{where } \tau_n = \text{round}\left(\frac{2\sqrt{R_1^2 + \left(X_1 - V(n-1)T_{p\text{-}p}\right)^2}}{c \times t_s} \right). \tag{15}$$

The process of sampling and summing is carried out sequentially and systematically for every pixel as shown in Fig. 6.26. The algorithm for implementing the sample-and sum operation is also provided in Fig. 6.27.

6.4.3   Simulation results

Based on the parameter values given in Tables 6.1 and 6.2, the two-dimensional plot of the target is then obtained as shown in Fig. 6.28.

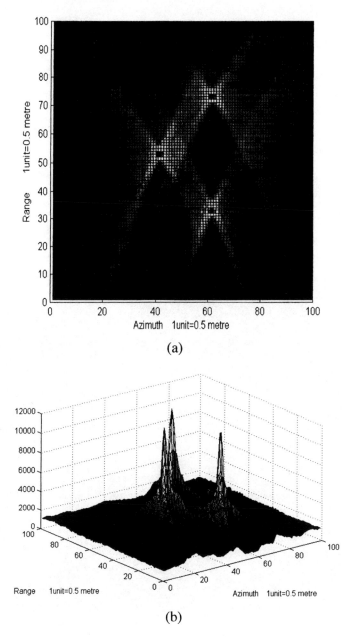

(a)

(b)

Figure 6.28: Reconstructed two-dimensional image. (a) Intensity plot of the target with three scattering points. (b) Three-dimensional plot of the target with three scattering points.

The three small patches indicate the location of the scattering points. With the two-dimensional plot above, the relative range and azimuth positions of the point scatterers on the target can be easily seen. The arrangement of the point scatterers used in the simulation is also shown in Fig. 6.29 for comparison.

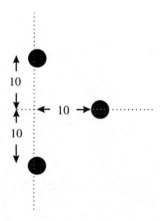

Figure 6.29: Geometry of three point-scatterers.

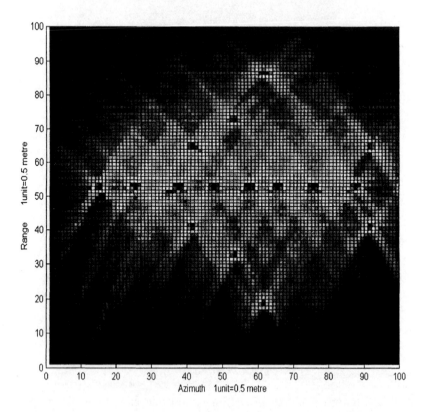

Figure 6.30: Reconstruction of an aircraft image; image of an aircraft using 16 point-scatterers.

As shown in Fig. 6.28, the locations of three points reconstructed via the procedure outlined in Section 6.4.2, match quite well with their actual position. The shape of the targets, however, has been distorted due to the sidelobes associated with the procedure. The white streaks surrounding each point target result from these sidelobes.

Next, a 16-point-scatterer model of an aircraft is considered. Each wing of the aircraft was considered to be made from two point-scatter centers. Each of the two drop tanks were replaced by a scatter center. The rest of the aircraft was considered to consist of 10 point-scatter centers. The ISAR reconstruction of Next, the three-point model is extended to a 16-point-scatterer model of an aircraft image, is shown in Fig. 6.30.

6.5 Imaging with line elements

In this section, the feasibility of applying the imaging process for point scattering to line scattering is studied. The purpose of imaging with line scattering is to reduce the computational burden imposed in point scattering imaging as a group of closely spaced point scatterers can be represented by only one line.

The general mathematical equation of the EM signal back-scattered by a line scatterer is given in eqn (108) of Chapter 3. For a chirp (linear FM) radar signal, the electric field is

$$E_p(r,z,t) = \hat{u}_h \frac{\mu_0}{4\pi} \left[\frac{r+L\cos\theta}{\sqrt{(r+L\cos\theta)^2 + z^2}} - \frac{r}{\sqrt{r^2+z^2}} \right] \frac{d}{dt}\left(I_0 e^{j\left(\omega(t-2R/c)+\pi a(t-2R/c)^2\right)} \right)$$

$$= \hat{u}_h \frac{\mu_0}{4\pi} \left[\frac{r+L\cos\theta}{\sqrt{(r+L\cos\theta)^2 + z^2}} - \frac{r}{\sqrt{r^2+z^2}} \right]$$
$$\times j\left(2\pi f_c + 2\pi a(t-\tau)\right)\left(I_0 e^{j\left(2\pi f_c(t-\tau)+\pi a(t-\tau)^2\right)} \right). \tag{16}$$

Therefore, the signal after mixing down is

$$E_p = \frac{\mu_0 I_0}{4\pi} \left[\frac{r+L\cos\theta}{\sqrt{(r+L\cos\theta)^2 + z^2}} - \frac{r}{\sqrt{r^2+z^2}} \right]\left(2\pi f_c + 2\pi a(t-\tau)\right)$$
$$\times \cos\left(2\pi\left(-f_c\tau + f_{if}t + 0.5a(t-\tau)^2\right)\right) \quad \text{for } \tau \le t \le T+\tau, \tag{17}$$

where

$$\tau = \frac{2\times\sqrt{z^2 + (r+(L\cos\theta)/2)^2}}{c}.$$

Comparing eqn (16) with eqn (1), it can be seen that the equations are different in the expressions for the time delay τ. In point scattering,

$$\tau = \frac{2 \times R}{c},$$

where R is the range of the radar to the point scatterer as shown in Fig. 6.31(a). However, in line scattering,

$$\tau = \frac{2 \times \sqrt{z^2 + (r + (L\cos\theta)/2)^2}}{c},$$

where the square root term defines the range of the radar to the centroid of the line as shown in Fig. 6.31(b).

Thus, in line scattering the time delay between the transmitted pulse and the return pulse gives the range of the centroid of the line and this center point on the line is imaged as shown in Fig. 6.32.

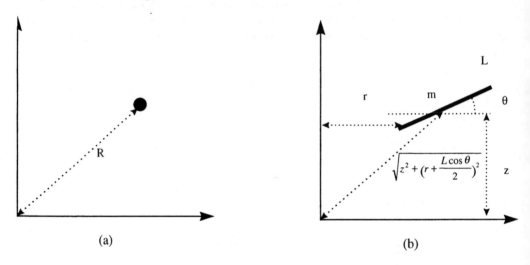

(a)   (b)

Figure 6.31: (a) Point scattering. (b) Line scattering.

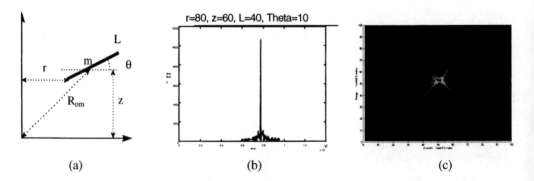

(a)   (b)   (c)

Figure 6.32: (a) Imaging geometry. (b) One-dimensional plot. (c) Two-dimensional plot.

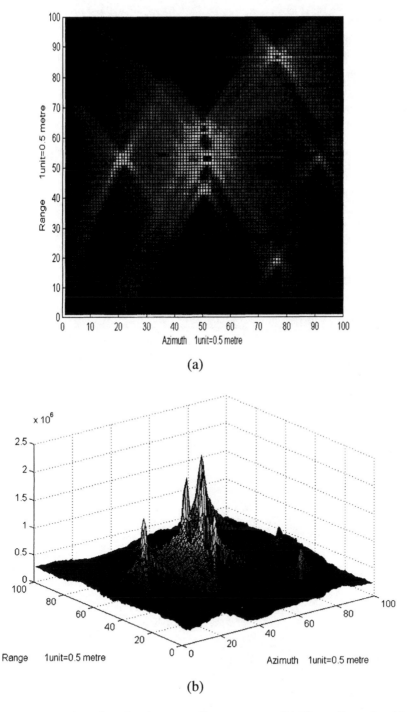

(a)

(b)

Figure 6.33: (a) Image of an aircraft using seven line-scatterers. (b) Three-dimensional image of an aircraft using seven line-scatterers.

For line imaging of the aircraft, seven finite length line-scatterers are used to model the aircraft scatter centers, instead of the 16 point-scatterers.

In this case, a few scattering points can be grouped as one finite line scatterer. The advantage is the reduction in the processing time with the line model compared to processing several points in the point model. Figure 6.33 shows the ISAR reconstructed image of the aircraft using a seven-line-scatterer model. It is observed that a general picture of the aircraft image emerges.

7 Smart antennas: mobile station antenna location

Stetson Oh Kok Leong, Ng Kim Chong, P.R.P. Hoole and E. Gunawan

7.1 Generations of mobile systems

Mobile cellular communication systems are dependent on high capacity, long-range radio technology. Considerable progress has been made to increase the capacity of the cellular network through advanced communication software and new hardware technology. Different multiple access techniques such as TDMA, FDMA and CDMA have been developed over the years to cope with the high capacity demand. All these multiple access techniques operate on the time-domain definition of the signal. More recently there has been considerable interest in operating on the signal in space-domain, leading to the idea of space division multiple access (SDMA) technique. All four multiple-access technologies seek:

- Voice, data, and video service that matches wired service quality.
- To cover the entire geographical area.
- To minimize equipment cost and size at the subscriber (mobile) and fixed (base) stations.

With low cost frequency synthesizers providing a clean frequency spectrum and microprocessor based control providing fast accurate switching and signal processing facilities, it is now possible to handle hundreds of subscribers in one cell of the large cellular network.

7.1.1 First-generation mobile system: FDMA

The advanced mobile phone system (AMPS), developed by the Bell Labs in the United States, was one of the first cellular, mobile communication system. Honeycomb-shaped cells covering a vast area with many base stations (BSs) are used in the AMPS systems. The BS of each hexagonal-shaped cell has a powerful transmitter and receiver, control unit and dipole antennas made directional by placing corner reflectors behind the antennas associated with each sector. The cells are arranged in such a way that, while covering a wide geographical area, there is no overlap. A subscriber or mobile station (MS) is made from a low power (less than or equal to 1 W) transmitter/receiver, a microprocessor-based control unit and normally a dipole antenna. As the MS roams through these cells during a call, control is handed over to the respective cell to receive signals from the MS as well as to transmit signals from the BS to the MS. The handing over of control from one cell to another is termed handoff. Identical frequencies are used in cells that are separated by one or two cells. Hence for every other cell, we use the same frequency band for communication, thus greatly increasing frequency reuse. Such frequency reuse enables the cellular system to serve an increased number of subscribers without demanding more space on the frequency spectrum.

Several other analog cellular systems based on the AMPS system paved the way for many other analog cellular systems. Amongst these newer first-generation systems are the total access communication system (TACS) and the Nordic mobile telephone (NMT) system in Europe. Analog frequency division multiple access (FDMA) technique was used for sharing the carrier frequency and time resources amongst many subscribers within each cell. In FDMA a single channel is allocated to one user at a time. This carrier frequency will be entirely dedicated to the single user until his call is completed or he has moved to another cell and has been handed over to another BS.

These first-generation analog systems are still widely in use in most countries. Upgrading these systems to improve on capacity and quality of service is an area of development. SDMA using smart antenna is one of the current technologies studied and its satisfactory operational deployment would bring about enormous changes to how we think and design cellular networks.

7.1.2   Second-generation mobile system: TDMA\CDMA

The cellular industry has been growing at a rapid rate. In the United States, for instance, in 1993 the number of wireless subscribers was about 10 million; since then it has exponentially grown to about 70 million in 1999. A growth rate of about 30% is expected in developed nations, and over 60% per year in developing countries. To meet the increasing demand, and thus the need for higher systems capacity and better link quality, digital communication systems are now used. The time division multiple access (TDMA) and code division multiple access (CDMA) are the two competing and yet incompatible digital standards that exist at present.

The most widely known TDMA digital standards are the North American Digital to using the FDMA protocol, it allows for time sharing as well. A single subs it switches criber is allocated to a specific carrier frequency for a short time. Subsequently another subscriber to the same frequency for another time slot, and so on. By going through cycles of this limited time period, in the order of milliseconds, several users may use a single Cellular (IS-54) and Global System for Mobile Communications (GSM). The TDMA is related to the FDMA in that, in addition carrier frequency.

The alternative to the TDMA digital standard is the CDMA standard (IS-95) which has been adapted by Telecommunications Industry Association (TIA) for the North American cellular telephone system. The CDMA system has been seen as a superior system to the TDMA systems, providing increased system capacity and better service quality. CDMA evolved from the spectrum technology that was popular in military circles at one time. In this digital spread spectrum technology, the signal containing the information is deliberately spread over the entire frequency band allocated for cellular communications and hence transmitted over a much wider bandwidth than that required for point-to-point communication at the same data rate. Subscribers are identified by unique digital codes spread over the entire frequency band. This is quite different to using separate RF carrier frequencies and bandwidths uniquely allocated for each user over a given period of time as in the case of TDMA. These digital codes are called pseudo-random code sequences, and are known by both the mobile station and the base station. These pseudo-noise (PN) codes permit a call from a particular user to be readily identifiable from all other calls concurrently existing in the vast frequency band. Each cell may have three or more sectors resulting from the use of sectored antennas. The entire frequency band may be reused in every sector of every cell. The major benefits of CDMA technology include a significant increase in system capacity increase, improved link quality, better coverage characteristics and enhanced privacy (see Fig. 7.1).

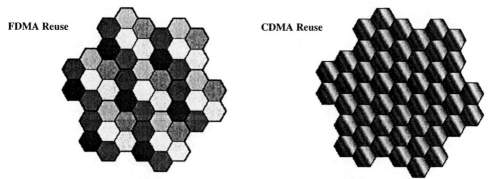

Figure 7.1: Cellular frequency reuse patterns.

Amongst the second-generation systems of the mid-1990s, the following are currently in use: (a) *GSM DCS-1900*. It uses the TDMA multi-access technology, and has uplink frequencies of 890–915 MHz (Europe) or 1850–1910 MHz (US) and downlink frequencies of 935–960 MHz (Europe) or 1930–1990 MHz (US). Carrier separation is 200 kHz and the GMSK (Gaussian minimum shift keying) modulation technique is used. (b) *CDMA-ONE IS95*. The uplink frequencies are 824–849 MHz (US Cellular) or 1850–1910 MHz (US PCS) and the downlink frequencies are 869–894 MHz or 1930–1990 MHz. The CDMA technology, the BPSK (binary phase shift keying) modulation and 1.25 MHz carrier separation are used. The NADC IS-54/IS-136 system using TDMA multi-access technology, $\pi/4$-DQPSK (differential quadrature phase shift keying) modulation and 30 kHz carrier separation also operates in the same frequency bands. (c) *PACS*. The uplink and downlink frequencies are 1850–1910 MHz and 1930–1990 MHz, respectively. It uses TDMA and $\pi/4$-DQPSK modulation with a 300 kHz carrier separation.

One aspect of the CDMA system that receives much attention is the near–far effect. If multiple subscribers or mobile stations (MSs) are all transmitting at the same signal power level to the base station, those nearest to the BS will tend to saturate the whole frequency spectrum that is being used. This will prevent those subscribers far away in the cell having a good enough link quality. This is the near–far effect, and is overcome by controlling the power of the MS (power control every 800 times each second) to ensure that the power received from all MSs, irrespective of distance, is almost equal. In open-loop power control, the downlink (BS to MS) path loss is used to determine the power at which the MS should transmit. In closed-loop power control, the BS observes the power level of signals from a particular MS, and sends a signal back to the MS to instruct it to raise or lower its transmit power. An alternative way to keep the MS power levels at the BS almost the same is to use distance dependent modulation techniques. In such an adaptive modulation technique, quarter-rate QPSK, half-rate QPSK, QPSK and 16-QAM modulation techniques may be used, with the 16-QAM used for the MSs nearest to the BS and the quarter-rate QPSK used for those furthest away from the BS within a given cell. By using adaptive modulation the signal power is controlled to prevent the near–far effect degrading the link quality.

7.2   Mobile radio environment

Objects surrounding the BS and MS severely affect the propagation characteristics of the uplink and downlink channels of cellular systems. This propagation path loss, including reflection and shadowing, tends to degrade system capacity. The height of the MS antenna (e.g. 2 m) is normally much lower than that of the surrounding buildings and natural features. Furthermore, the carrier frequency wavelength is also much less than the size of the surrounding structures. Due to this, an MS will experience significant changes of its received signal strength as it moves. The mobile receiver is characterized by "multipath reception". Its received signal contains a number of electromagnetic waves from the same source (transmitter) arriving at the receiver antenna along different paths. Even when line-of-sight is available (this is rare in urban areas), the additional electromagnetic waves beside the direct wave result from the reflection, refraction, scattering and diffraction of the transmitted signal off objects along the propagation path. These extra waves arrive at the receiving antenna displaced with respect to each other in time and space. Due to this phenomenon, the resultant received signal that appears at the receiver amplifier could be much weaker or stronger than the direct wave. As shown in Section 2.11 every half wavelength in space, nulls of fluctuation at the baseband can be visualized, although the nulls do not occur at the same level. If the mobile station moves at a faster speed, the rate of fluctuation observed by the receiver becomes faster.

Fading due to multipaths and shadowing, Doppler spread (due to the motion of one of the antennas) and delay spread are some of the main channel effects that arise from these phenomena (see Fig. 7.2).

7.2.1   Fading

When a mobile station moves over small distances of a fraction of a wavelength, the instantaneous field strength at the receiver antenna may rapidly fluctuate as much as four or five orders of magnitude. This is known as signal fading, or small-scale fading. There are two approximately separable effects known as fast and slow fading. Fast fading is characterized by deep fades that occur within fractions of a wavelength and is caused by multipath scattering of the signal off objects in the vicinity of the mobile. It is most severe in heavily built-up areas where the number of waves arriving from different directions with different amplitudes and phases cause the signal amplitude to follow a Rayleigh distribution. However, fast (Rayleigh) fading also occurs to a certain extent in suburban areas.

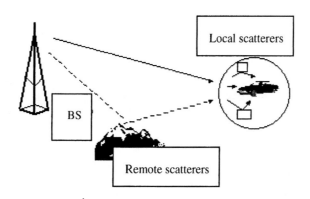

Figure 7.2: Multipath.

The overall signal with multipath reflections can be expressed as (Razavi, 1999)

$$x_R(t) = a_1(t)\cos(w_ct + \theta_1) + a_2(t)\cos(w_ct + \theta_2) + \cdots + a_n\cos(w_ct + \theta_n) \qquad (1)$$

$$= \left[\sum_{j=1}^{n} a_j(t)\cos\theta_j\right]\cos w_ct - \left[\sum_{j=1}^{n} a_j(t)\sin\theta_j\right]\sin w_ct . \qquad (2)$$

For a large number of multipaths (i.e. large n), each summation has a Gaussian distribution. Denoting the first summation by A and the second by B gives

$$x_R(t) = \sqrt{A^2 + B^2}\,\cos(w_ct + \phi), \qquad (3)$$

where $\phi = \tan^{-1}(B/A)$.

For such a scenario of signals, it can be shown that the envelope of the received signal has a Rayleigh density function given below:

$$p(y) = \begin{cases} \dfrac{y}{\sigma^2} l^{-y^2/2\sigma^2}, & y \ge 0, \\ 0, & y < 0, \end{cases} \qquad (4)$$

where y is a random variable representing the signal level fluctuation and σ is its standard deviation.

When there is a direct line-of-sight component present, as would be the case when the MS is traveling along a highway or open countryside, together with the Rayleigh or multipath, the received signal becomes Rician. Its probability density function is as follows:

$$p(y) = \begin{cases} \dfrac{y}{\sigma^2} e^{-(y^2+s^2)/2\sigma^2} J_0\left(\dfrac{ys}{\sigma^2}\right), & y \geq 0, \\ 0, & y < 0, \end{cases} \tag{5}$$

where $J_0(.)$ is the modified Bessel function of the zeroth order and s^2 is the mean power of the direct path.

Slow fading related to shadowing is the result of buildings and trees that stand between the BS and MS antennas. It is observed that such slow fading exhibits a log-normal distribution given by

$$p(z) = \begin{cases} \dfrac{1}{\sqrt{\pi}\sigma z} e^{(\log z - \mu)^2 / 2\sigma^2}, & z > 0, \\ 0, & \end{cases} \tag{6}$$

where μ is the mean of the random variable z and the standard deviation is between 5 and 10 dB.

7.2.2 Doppler spread

In a wireless channel with multipath Rayleigh fading signals, let the nth reflected wave arrive from an angle θ_n relative to the direction of the motion of the MS antenna. We have shown in Section 2.3 that the frequency of the received signal will go through a shift due to the motion of the antenna. The change in the received signal frequency of this Rayleigh fading signal is known as the Doppler shift. It can be represented by the formula

$$\Delta f_n = \frac{v}{\lambda} \cos \theta_n \tag{7}$$

where v is the speed of the mobile antenna and λ is the wavelength.

Hence motion of the MS antenna that results in the Doppler shift produces phase shifts of each reflected wave. When the waves arrive at the antenna, since they all have different phase shifts, the amplitude of the resulting composite signal will be modified by the Doppler shift, and hence the velocity vector of the MS.

When the number of multipath waves that arrive at the receiver is large and each wave has its own random angle of arrival (hence with its own Doppler shift), the baseband power spectrum of the vertical electrical field has the following form:

$$S(f) = \frac{3\sigma}{2\pi f_m}\left[1 - \left(\frac{f - f_c}{f_m}\right)^2\right]^{-1/2}, \quad f_c - f_m < f < f_c + f_m, \tag{8}$$

where the angle of arrival is assumed to be uniformly distributed within 0 to 2π, f_m = v/λ is the maximum Doppler shift, f_c is the carrier frequency, and σ is the mean signal power. It is seen from eqn (8) that the received signal strength will be dependent on the Doppler shift and hence the velocity vector of the MS.

Doppler spread causes time-selective fading since it directly impinges on the phase of the signals. Thus the instantaneous received signal amplitude varies with time. The signal amplitude in the presence of Doppler fading is characterized by and inversely proportional to the coherence time. We define the coherence time as the time separation over which the channel impulse response at two time instants remains strongly correlated.

7.2.3   Delay station spread

Delay spread is a well-known phenomenon at high frequency electromagnetics. In wired systems we get delay spread due to the high frequency portion of a signal (e.g. a rectangular pulse) traveling much faster than the lower frequency portion of the signal. Hence at the line termination the signal looks spread out in the time domain, since the lower frequency energy arrives later than the high frequency energy. In wireless systems, time spread occurs due to multipaths, in other words due to reflection and differences in the distance traveled by each reflected signal. When the distances are different, although the signals travel at the velocity of light, their arrival times will be different and proportional to the distance traveled. Hence multipath effects create time dispersion and a spreading of the signal. Urban delay spreads of around 3 μs are commonplace. Delay spread results in frequency-selective fading, that is, it is dependent on the frequency. If the variation of the delay is comparable with the symbol period, delay signals from an earlier symbol may interfere with the next symbol, causing intersymbol interference (ISI). It can be characterized by the coherence bandwidth, which is inversely proportional to the delay spread. Coherence bandwidth is defined as the maximum frequency difference for which two frequency-shifted signals are still strongly correlated in terms of either amplitudes or phases.

7.3   Mobile station positioning

The importance and usefulness of accurate MS position information has been recognized in recent years. Besides E911 services, it opens up a wide area of possibilities. For example, it can play an important role on improving the system capacity either directly or indirectly, ranging from the development of better beamforming algorithm and enhanced adaptive modulation technique to effective cellular system design.

A hierarchical cell system using the position and velocity of the MS offers a good compromise between an efficient use of available channels while simultaneously keeping the number of handoffs small. In this system, cells of different sizes coexist. Equal-sized cells are grouped into layers, which overlay on top of one another to form a hierarchy. Thus, a layer with large cells can be assigned to fast moving MSs, while a layer with small cells or microcells to the slow moving MSs. This serves two purposes:

(1) The cells of small and large radius provide a more economically efficient system for higher and lower traffic densities. (2) Subscribers of lower and higher mobility can efficiently be provided with service in the small cells and umbrella cells, respectively.

MS position estimation can be categorized under MS-based and BS-based position-finding techniques. For MS-based techniques, the MS makes use of signals transmitted by the surrounding structures to compute its own position. These structures can be BSs in a cellular network or satellite network in the sky. The latter technique relies solely on the existing infrastructure of cellular BSs.

7.3.1   Global positioning satellite (GPS)

A medium-earth-orbit (MEO) satellite system having about 24 satellites is used in the GPS system (see Fig. 7.3). A GPS receiver installed at the MS, as in a taxi for instance, provides the self-position estimate of the MS using a group of satellites in the MEO network. Each satellite transmits a spread spectrum signal to earth on the L-band (centered at 1575.42 MHz). Using an accurate clock for precise timing, the GPS receiver measures the time delay between the signals leaving, say, three satellites in the sight of the MS and arriving at the MS receiver. This allows calculation of the exact distance from the MS to each satellite. Thus the MS position can be computed using the triangular method, providing coordinates in latitude, longitude and altitude. In practice, a fourth satellite is used to correct receiver clock errors. GPS is one of the most popular radio navigation aids due to its high accuracy. A commercial GPS receiver costs under US\$ 150 with an accuracy of approximately 50 m.

However, GPS has certain disadvantages. The MS has to carry a GPS receiver, which means increased weight, size and battery drainage. Furthermore, when the L-band electromagnetic signals are blocked or attenuated by buildings, foliage and heavy cloud cover, the resulting estimate of the MS position may be quite inaccurate due to missed data. One further technical problem is to do with the antenna: the MS antenna is designed to operate in the UHF (0.3–1.0 GHz), L-band (1.0–2.0 GHz) or S-band (2.0–4.0 GHz) wireless communication frequencies, whereas the satellite GPS system operates in the L-band. Where the frequencies differ, antenna performance will also be affected.

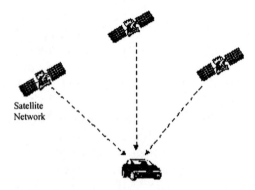

Figure 7.3: Global positioning satellite.

7.3.2 MS positioning in the cellular network

7.3.2.1 BS-based positioning

In smart antenna technology, BS has played a critical role in making it possible to install large array antennas to achieve adaptive, steerable beamforming where the multiple beams keep track of the MS. With a smart antenna with 10 elements installed at the BS, in the uplink an SNR improvement of 10 dB, a range increase of 2, an SINR improvement of 20 dB and a data rate increase of 2 could be achieved. In the downlink the BS emission can be reduced by 90% and a capacity increase of about 3 is also possible. However, in order to generate these smart beams, the position of each MS should be located. BS-based positioning is implemented mostly using a triangular method based on signal strength, angle of arrival, time of arrival measurement or their combination.

Mobile position estimation using signal strength measurement is one of the most well-known methods using the path loss attenuation with distance information. Its primary source of error is multipath fading and shadowing. Variations in the signal strength can be as great as 30–40 dB over a distance of the order of half a wavelength. Signal strength averaging can help, but low-mobility MSs may not be able to average out the effects of multipath fading, and there will still be adverse effects due to shadow fading. The errors due to shadow fading may be handled by using pre-measured signal strength contours centered at the BSs. However, this approach requires that the contours be mapped out for each BS and a data bank be available for use.

This signal strength based approach to position estimation is difficult in digital systems where transmission power is controlled close to 1000 times every second. The CDMA system employs power control in order to combat the near–far effect. The TDMA system uses power control to conserve battery power of the MSs in the upward link. Thus, for such systems, unless the instantaneous transmission power of the MSs is known it may not be possible to achieve reasonable accuracy.

The angle-of-arrival (AOA) methods are sometimes also referred to as direction-of-arrival (DOA) methods. The AOAs of the signal from the MS are calculated at the BS by using adaptive phased antenna arrays at the BS. The main beam of the antenna array is electronically steered until it locks on to the signal arriving at the MS. Two closely spaced antenna arrays are used to scan horizontally to get the exact direction of the peak strength of the incoming MS signal. Since the antenna elements are closely placed, the time delay between the two elements seen by a signal as it propagates across the array may be modeled by a phase shift. This is the basic arrangement for most AOA estimation algorithms. Accurate AOA is estimated when the signal coming in from the MS is a direct line-of sight (LOS) signal. If the antenna array beam should lock on to a reflected signal the AOA estimate will obviously be wrong. Furthermore, the AOA estimator performance degrades as the distance between the MS and the BS increases.

The time-of-arrival (TOA) based estimation of position is also possible. In free space the time taken for an electromagnetic wave to travel over a distance is proportional to d, since the velocity of the wave is equal to the constant speed of light. Hence the BS may determine the distance d by first indirectly determining the time that the signal takes to travel from the source to the receiver on the forward or the reverse link. This TOA may be obtained by measuring the time in which the MS responds to an inquiry or an instruction transmitted to the MS from the BS. The total time elapsed from the instant the command is transmitted to the instant the MS responds may be stored, and this time is equal to the sum of the round trip signal delay and any processing and response delay within the MS unit. If the MS microprocessor processing delay for the desired response is known with sufficient accuracy, it can be subtracted from total measured time. The resultant time would give the total round trip delay. Half of the round trip delay would be an estimate of the signal delay in one direction. If we multiply this with the velocity of light, we will get the distance between the BS and the MS. If the MS signal can be detected at a minimum of three BSs, then the MS position can be computed by the triangulation method.

There are certain problems with the TOA method. The estimate of the microprocessor processing time and response delay of MS electronics within the MS may be difficult to determine in practice. Indeed different manufacturers of MS phones will use different electronics and circuitry, giving rise to quite different microprocessor-discrete electronics response times. Furthermore LOS between the BS and MS is not available, and severe timing errors could occur. The problem will be difficult to handle when there are many multipath signals arriving at the BS.

7.3.2.2   MS-based positioning

This method requires the MS to use signals from several BSs to calculate its position. It is a self-positioning system, where the MS estimates its own position. Again the signal strength (this time from BS to MS) or the TOA information may be used. However, modification to the existing MS software will be necessary since the position computation will be performed at the MS. The control microprocessor must carry the position-estimation software in order to store and process information from four or more BSs (see Fig. 7.4).

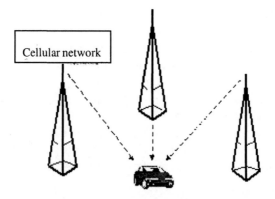

Figure 7.4: MS positioning in the cellular network.

The MS receiver must be capable of simultaneously processing information from at least four BS frequency channels, either for signal strength or time delay information.

Data from three BSs are required for the position estimation, while the fourth is needed to cycle through all available signals to ensure the best three BS signals are being used for the position estimation. In practical implementation, it is necessary for the network to maintain a BS location database and transfer information to the MS whenever it is requested.

7.4   Position and velocity estimation in cellular systems

Information about the position and velocity of mobile stations (MSs) can be obtained using two available quantities (Stuber, 1996), namely the signal strengths of the MS at different base stations (BSs) and the corresponding propagation times. However, both quantities are subject to strong fluctuations caused by short-term fading, shadowing and reflections, which make them useless unless a sophisticated method is used to counter this problem and translate the quantities into the required information.

A new method of estimating the position and velocity of an MS in a cellular network, based on an electric-field strength model, is presented in this chapter. The algorithm developed uses the principle of maximum likelihood estimation and is tested in MATLAB^{TM} simulation experiments for different channel conditions.

7.4.1   Antenna signal model

In this section, a model for electromagnetic fields radiated by an infinitesimal element carrying current of any geometrical shape is presented. The geometry is defined in Fig. 7.5.

From the radiated electric fields from a finite-sized wire antenna obtained in Chapter 3, we get

$$E(R,t) = \mathbf{u}_R \frac{h\cos\theta}{4\pi}\left[\left(\frac{\mu_0}{\varepsilon_0}\right)^{1/2}\frac{2[I]}{R^2}+\frac{2[Q]}{\varepsilon_0 R^3}\right]+\mathbf{u}_\theta\left[\frac{h\sin\theta}{4\pi}\frac{\mu_0}{R}\frac{d[I]}{dt}+\left(\frac{\mu_0[I]}{\varepsilon_0 R^2}\right)^{1/2}+\frac{[Q]}{\varepsilon_0 R^3}\right].$$

(9)

In cylindrical coordinates, using cylindrical coordinate unit vectors $\mathbf{u}_r$ and $\mathbf{u}_z$, eqn (9) may be written as

$$E(r,z,t) = \mathbf{u}_r\left[\frac{3h}{4\pi}\left(\frac{\mu_0}{\varepsilon_0}\right)^{1/2}\frac{rz}{(r^2+z^2)^2}[I]+\frac{\mu_0 h}{4\pi}\frac{rz}{(r^2+z^2)^{2.5}}\frac{d[I]}{dt}+\frac{3h}{4\pi\varepsilon_0}\frac{rz}{(r^2+z^2)^{2.5}}[Q]\right]$$
$$+\mathbf{u}_z\left[\frac{h}{4\pi}\left(\frac{\mu_0}{\varepsilon_0}\right)^{1/2}\left(\frac{2z^2}{(r^2+z^2)^2}-\frac{r^2}{(r^2+z^2)^2}\right)[I]-\frac{\mu_0 h}{4\pi}\frac{r^2}{(r^2+z^2)^2}\frac{d[I]}{dt}\right.$$
$$\left.+\frac{h}{4\pi\varepsilon_0}\left(\frac{z^2}{(r^2+z^2)^{2.5}}-\frac{r^2}{(r^2+z^2)^{2.5}}\right)[Q]\right].$$

(10)

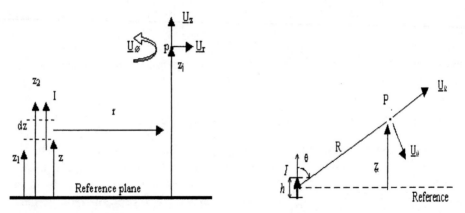

Figure 7.5: An infinitesimal current carrying element and its coordinate system.

It can be further simplified as the receiving antenna will only pick up the E_z, the vertical polarized signal. Therefore, by ignoring E_r and assuming that $[Q]=0$,

$$E'(r,z,t) = \mathbf{u}_z \left[\frac{h}{4\pi} \left(\frac{\mu_0}{\varepsilon_0} \right)^{1/2} \frac{2z^2}{(r^2+z^2)^2} - \frac{r^2}{(r^2+z^2)^2} [I] - \frac{\mu_0 h}{4\pi} \frac{r^2}{(r^2+z^2)^2} \frac{d[I]}{dt} \right]. \quad (11)$$

By setting $h = dz$, $z = z_j - z$, $dz = -dz$ and integrating eqn (11), and considering only far-field regions, the electric field strength radiated by a linear dipole antenna of finite length $(z_2 - z_1)$ is of the form

$$E(r,z,t) = \frac{\mu_0}{4\pi} \left[\frac{z_j - z_1}{\sqrt{r^2 + (z_j - z_1)^2}} - \frac{z_j - z_2}{\sqrt{r^2 + (z_j - z_2)^2}} \right] \frac{d[I]}{dt}, \quad (12)$$

$$[I] = \text{Re}\left(I_0 e^{jw(t-R/c)} \right), \quad (13)$$

where I_0 is the amplitude and ω is the frequency of the current flowing through the transmitter antenna. Thus,

$$\frac{d[I]}{dt} = j\omega I_0 e^{j\omega(t-R/c)}. \quad (14)$$

The signal processor operates only on the magnitude of the measured electric field strength. Hence eqn (14) is reduced to

$$\left| \frac{d[I]}{dt} \right| = \omega I_0. \quad (15)$$

Therefore, eqn (12) can be expressed as

$$E(r,z) = A_0 \left[\frac{z_j - z_1}{\sqrt{r^2 + (z_j - z_1)^2}} - \frac{z_j - z_2}{\sqrt{r^2 + (z_j - z_2)^2}} \right], \quad (16)$$

where $A_0 = \mu_0 \omega I_0 / 4\pi = 10^{-7} \omega I_0$ and

$$I_0 = \sqrt{\frac{P_r}{R_r}}. \quad (17)$$

P_r is the power radiated and R_r is the radiation resistance of the transmitter antenna. The parameters of the model are defined in Fig. 7.6. The advantage of using this model is that it provides a more realistic picture of a BS–MS communication link as it takes into account the distance r between the BS and the MS as well as their respective heights (z_1, z_2 and z_j) with respect to the ground.

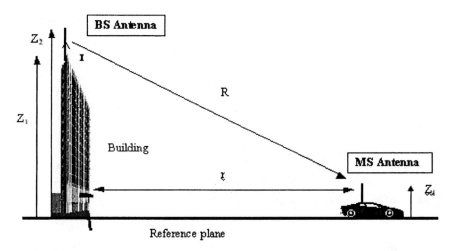

Figure 7.6: E-field model parameters.

7.4.2 Position and velocity estimation (PVE) algorithm

The measured instantaneous electric field strengths E_m in the presence of additive white Gaussian (AWG) noise is modeled by

$$E_m = E + n, \quad (18)$$

where E is defined in eqn (16) and n is the AWG noise. Since E is a non-linear function of r, it can be expressed as

$$E(r) = E(r_0) + E'(r_0)\,\Delta r + 0.5\,E''(r_0)\Delta r^2 + \cdots . \quad (19)$$

Ignoring higher order terms and writing $E_0 = E(r_0)$, let

$$J_0 = E'(r_0) = \frac{\delta E_0}{\delta r_0}, \tag{20}$$

$$E(r) = E_0 + J_0 \Delta r. \tag{21}$$

Substituting eqn (21) into eqn (18) gives

$$n = (E_m - E_0) - J_0 \Delta r = \Delta z - J_0 \Delta r \tag{22}$$

We seek to minimize

$$U(\Delta r) = n^T n. \tag{23}$$

Substitute for n using eqns (21) and (22), and let

$$\frac{\partial U(\Delta r)}{\partial r} = 0,$$

and we get

$$\Delta r = (J_0^T J_0)^{-1} J_0^T \Delta z. \tag{24}$$

Therefore, r can be estimated using a maximum likelihood estimator (MLE) of the following form:

$$r_{n+1} = r_n + (Jr_n^T Jr_n)^{-1} Jr_n^T (E_m - E_n), \tag{25}$$

where E_m is a column vector that represents the measured electric field strengths at the BSs and Jr_n is the Jacobian matrix of the electric field strengths E_n. The superscript T represents a non-conjugate transpose.

$$Jr_n = \frac{\partial E_n}{\partial r_n} = -A_0 \left[\frac{z_j - z_1}{[r_n^2 + (z_j - z_1)^2]^{3/2}} - \frac{z_j - z_2}{[r_n^2 - (z_j - z_2)^2]^{3/2}} \right] r_n. \tag{26}$$

Using the estimated r vector, which gives the distances between the BS's positions (PB) and the MS, an estimated position of the MS (PM) can be obtained. However, since the radii of the circles taken from the r estimates are unlikely to arrive at an exact cross intersection, further processing is necessary to reach a closer estimate of the actual position of MS.

Thus, an iteration equation using MLE similar to eqn (25) is formed. We have

$$PM_{k+1} = PM_k + (Jp_k^T Jp_k)^{-1} Jp_k^T (0 - errp_k), \tag{27}$$

where $errp = |PM - PB| - r$. Jp_k is defined as

$$Jp_k = \begin{bmatrix} \partial errp_1/\partial x & \partial errp_1/\partial y \\ \partial errp_2/\partial x & \partial errp_2/\partial y \\ \partial errp_3/\partial x & \partial errp_3/\partial y \end{bmatrix}. \tag{28}$$

The velocity of the MS (V) can be obtained from sequential PM estimates as follows:

$$V_m = (PM_m - PM_{m-1})/T, \tag{29}$$

where T is the time interval between two discrete positions of the MS.

7.4.3 Simulation scenario

The PVE algorithm is tested by running simulation in MATLAB. The area of interest is a region of 6000×6000 m that contains a total of 16 cells, each with a cell radius of 1000 m. A BS, indicated by the alphabet "A" shown in Fig. 7.7a, is located at the center of each cell. A MS travels along a route indicated by the dotted line with its velocities shown in Fig. 7.7b. Twenty sampling points are taken along the route at a time interval of 96 ms. The actual positions of the MS are marked by "x".

As the MS travels along the route, it establishes a wireless communication link with the base station within the cell. This BS is referred to as the control base station (CBS). It is assumed that the desired signal always arrives from the CBS. At least three BSs, including the CBS, are required to compute the position of the MS:

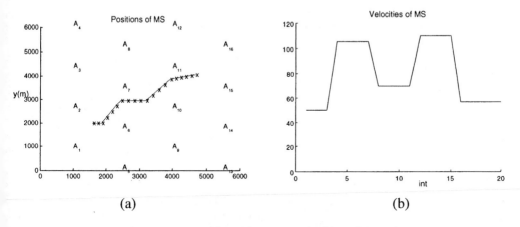

(a)   (b)

Figure 7.7: Actual positions and velocities of the MS.

The assumptions made in computing the BS position and velocity are as follows:

(1) The network maintains a database of the locations of the BSs and this is sent to the MSs whenever it is requested.
(2) Line-of-sight (LOS) is available between the MS and the BSs or a strong direct signal path exists.
(3) The control base station used is the BS closest to the MS.

(4) No transmission power control in the forward link (downward link) is used for the signal being used for estimation. For example in CDMA, the pilot tone may be used for position and velocity estimation.

These assumptions are made to simplify the simulation scenario in order to study the performance of the PVE algorithm within reasonable constraints. To further examine the performance of the algorithm, it is also tested in areas with smaller cells (e.g. radius = 200 m). Cell sizes with radii of 1000 m and 200 m are referred to as M-cells and P-cells, respectively (see Table 7.1).

Table 7.1: Simulation parameters.

Parameters	Description
$f = 900$ MHz	Signal frequency in GSM system
$c = 3 \times 10^8$ m/s	Speed of light
$z_1 = 60$ m	Height of the lower part of the BS antenna
$z_2 = z_1 + \lambda/2$	Height of the upper part of the BS antenna
$z_j = 1$ m	Height of the MS antenna
$R_r = 73\ \Omega$	Radiation resistance of the BS antenna
$P = 10$ W (M-cells)	Radiation power of the BS antenna
$P = 0.1$ W (P-cells)	

7.4.4   Channel models

The position–velocity estimator (PVE) algorithm is tested under different fading channel conditions. The channel models used in the simulation are described in the following subsections.

7.4.4.1   Additive white Gaussian (AWG)

A Gaussian distributed random generator is used to generate the required random number N_{AWG}, where m is the mean amplitude and σ is the variance of the noise. A_n is a scaling factor used to adjust to the required SNR. All channel models in the simulation assumed zero mean and a variance of 0.161 unless otherwise stated.

$$E_m = E + A_n N_{AWG}(m,\sigma),$$
(30)

$$SNR = 10 \lg \frac{E^2}{N_{AWG}^2}.$$
(31)

7.4.4.2 Rayleigh fading

The Rayleigh fading channel is simulated by (Loo, 1991):

$$N_{\text{Rayleigh}} = \sqrt{N_{\text{AWG}}(m,\sigma)^2 + N_{\text{AWG}}(m,\sigma)^2}, \tag{32}$$

$$E_m = N_{\text{Rayleigh}} E + A_n N_{\text{AWG}}(m,\sigma). \tag{33}$$

7.4.4.3 Dominant reflected path

To study the effects of multipath that results in the $1/r^4$ power decay, a dominant reflected path is assumed, with E_d and E_r the direct and reflected components, respectively (see Fig. 7.8). E_d is the E-field model defined in eqn (16). Assuming total reflection on the ground, the reflective coefficient R_c is -1. T_d is the time delay arrival of E_r at the MS with reference to E_d. k is a scaling factor calculated by $(1-\text{SNR}/100)$.

$$E_m = |E_d + kR_c E_r| + A_n N_{\text{AWG}}(m,\sigma), \tag{34}$$

$$E_r = A_0 e^{-j\omega T_d} \left[\frac{z_j - z_{1r}}{\sqrt{r^2 + (z_j - z_{1r})^2}} - \frac{z_j - z_{2r}}{\sqrt{r^2 + (z_j - z_{2r})^2}} \right], \tag{35}$$

$$T_d = (R_r - R_d)/c, \tag{36}$$

where

$$z_{1r} = z_1 + z_j,$$

$$z_{2r} = z_2 + z_j,$$

$$R_r = \sqrt{(z_2 + z_j)^2 + r^2},$$

$$R_d = \sqrt{(z_2 - z_j)^2 + r^2}.$$

7.4.4.4 Rician fading

When there is a direct line-of-sight component present, together with the Rayleigh or multipath, it becomes Rician (Loo, 1991). In this case

$$N_{\text{Rician}} = \sqrt{(E + N_{\text{AWG}}(m,\sigma)^2 + N_{\text{AWG}}(m,\sigma)^2} \tag{37}$$

$$E_m = N_{\text{Rician}} + A_n N_{\text{AWG}}(m,\sigma). \tag{38}$$

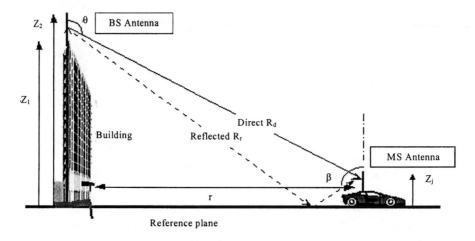

Figure 7.8: Dominant reflected path.

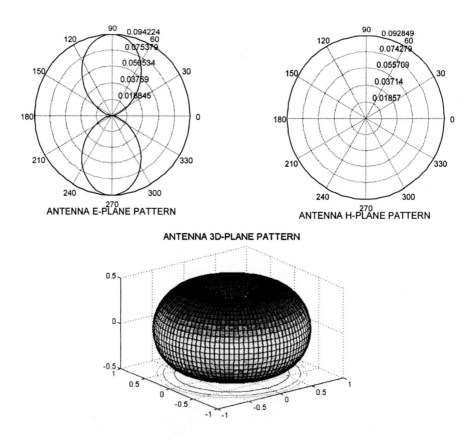

Figure 7.9: Radiation pattern of an infinitesimal wire antenna.

7.4.5   Antenna radiation pattern

All antennas for both BS and MS in the simulation are assumed to be omni-directional. The antenna patterns for E-plane (zy-plane), H-plane (xy-plane) and three-dimensional region (xyz-region) are shown in Fig. 7.9.

7.4.6   Initial values

The PVE algorithm requires an initial set of r values (distance from MS to each of the selected BSs) to proceed. These values can be obtained by making use of timing information provided by the networks. It is found that with the help of this timing information to generate the required initial values, the algorithm is able to produce accurate results.

In GSM systems (Fig. 7.10), useful timing information called time advance (TA) measurement is provided that gives a round trip propagation time for the microwave signal to travel between an MS and a particular BS. TA measurement is a technique used in GSM to inform an MS how much time in advance of the reference signal it should transmit in order to synchronize correctly at the BS. It reduces the guard period between time slots.

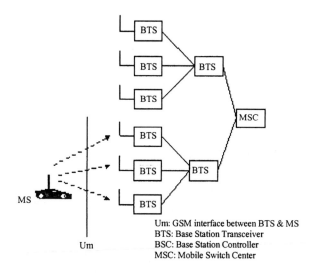

Figure 7.10: Simplified GSM network components.

The TA field is coded with eight bits allowing for 63 steps, where one step is one bit period (BP). BP, the fundamental unit of a frame, is equal to 48/13 µs (3.69 µs). The distance corresponding to a duration of BP is $r = cT_b/2 = 554$ m. This is twice the distance between MS and BS as the uplink is timed relative to the frame structure in the downlink. By using the respective TA measurement, the distance between an MS and BSs (r values) can be coarsely estimated. Although these values are not accurate enough to be used for the estimation of the MS's position, they are useful as the initial values of r.

In CDMA systems, similar timing information is also available, known as pilot strength measurement. This information is available and constantly updated at the MS. During each conversation, the MS continuously searches for new pilots, as well as the strengths of the pilots associated with the forward traffics channels. MS reports each pilot's strength and PN phase (or arrival time) to the control base station whenever required. The pilot arrival time is the time of occurrence, as measured at the MS antenna, of the earliest arriving usable multipath component of the pilot. Thus the reported pilot PN phase can be used to estimate the round trip delay to the BS from which the pilot is transmitted. The time delay information can then be used to compute a set of initial r values.

In the simulation scenario, the only time this information is used to provide the initial r values is when handoff occurred. In such cases, the strongest signal strength received by the MS is assumed to come from the closest BS (control base station). Hence, the handoff is based on some signal strength based handoff algorithms. At any other time (MS travels within a cell), the previous position of the MS is used instead to compute the initial r values.

7.4.7   E-field strength measurement

The GSM network provides signal strength measurement of different BSs measured at a mobile. The corresponding measured E-field strengths can be obtained by the following equation:

$$E_m = \sqrt{\frac{2(120\pi)P_m}{A_e}},$$ (39)

where P_m is the signal strength and A_e is the effective aperture of the antenna. The effective aperture of a half-wave dipole antenna is $0.13\lambda^2$, where λ is the wavelength. The studies are carried out for two typical frequencies used in mobile communications: 900 MHz and 1900 MHz.

7.4.8   Simulation results

The iteration begins with the initial values of r and proceeds until a convergence accuracy of 10^{-4}. It is found that the PVE algorithm can converge very fast, usually in less than eight iteration cycles. The average convergence cycle is about cycle 6. Figure 7.11 shows the convergence curves for the estimated r vector (r_1, r_2, r_3) for MS positions 1, 8 and 15 under AWGN channel for an SNR of 10 dB. The figure with the tabulated results in Table 7.2 show that the estimated r is quite accurate as compared to the actual r.

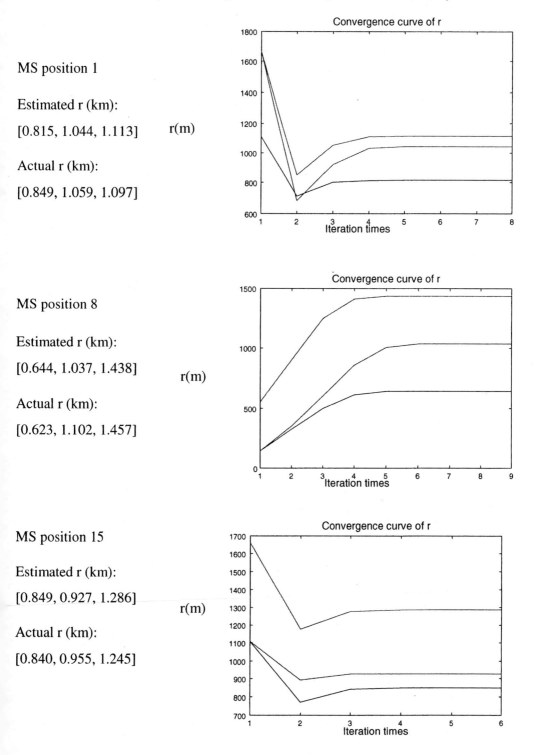

MS position 1

Estimated r (km):

[0.815, 1.044, 1.113]

Actual r (km):

[0.849, 1.059, 1.097]

MS position 8

Estimated r (km):

[0.644, 1.037, 1.438]

Actual r (km):

[0.623, 1.102, 1.457]

MS position 15

Estimated r (km):

[0.849, 0.927, 1.286]

Actual r (km):

[0.840, 0.955, 1.245]

Figure 7.11: Iteration cycle for *r* vector.

Table 7.2: Comparison of estimated *r* with actual *r*.

MS position	Estimated *r* vector (km)	Actual *r* vector (km)
Position 1	[0.815, 1.044, 1.113]	[0.849, 1.059, 1.097]
Position 8	[0.644, 1.037, 1.438]	[0.623, 1.102, 1.457]
Position 15	[0.849, 0.927, 1.286]	[0.840, 0.955, 1.25]

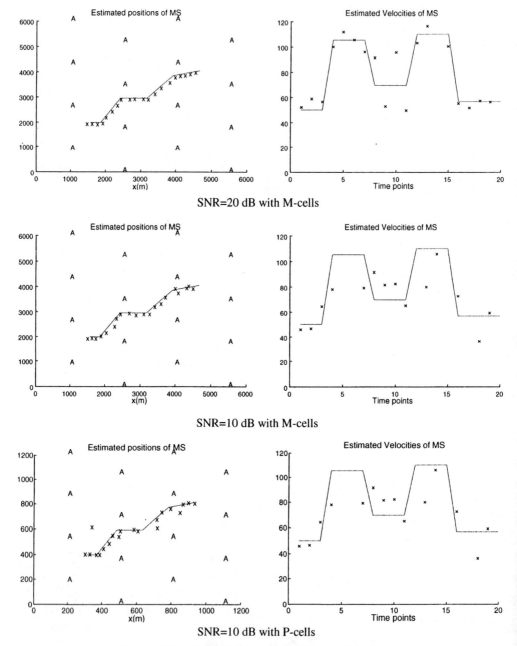

Figure 7.12: Simulation results for Rayleigh fading.

Simulation results for the estimation of MS position and velocity for the Rayleigh channel condition are shown in Fig. 7.12. Their respective average position and velocity errors are tabulated and shown in Table 7.3. The estimated MS position is marked by "x" while the actual route taken by the MS is indicated by the dotted line. The overall results shown in Table 7.3 and Fig. 7.12 confirmed the good accuracy of the PVE algorithm. Based on the models used, the average position error is largest when a dominant reflected path is considered. This may be caused by a few poor estimates. However, the error may be greatly reduced if a filtering technique such as Kalman filter is employed.

It is observed that the PVE algorithm performs better in an area with smaller cells (cell radius = 200 m). For example, under Rayleigh fading channel for an SNR of 10 dB, the average position error for an area with a cell radius of 1000 m is 68 m, which is about 66% higher than that of an area with smaller cells. However, since the E-field model is derived based on the assumption that the transmitter is in the far-field region, the model could not be used in an area with very small cell structure such as picocell, which has a radius of only a few meters. The E-field model used needs to be redeveloped to take into account the near-field region.

Table 7.3: Simulation results.

Channel model	SNR (dB)	Cell type	Average position error (m)	Average velocity error (km/h)
Gaussian	20	M-cell	21	8
	10	M-cell	63	23
	10	P-cell	20	32
Rayleigh	20	M-cell	33	14
	10	M-cell	68	29
	10	P-cell	23	36
Reflected path	20	M-cell	135	42
	10	M-cell	148	47
	10	P-cell	85	68
Rician	20	M-cell	29	13
	10	M-cell	69	29
	10	P-cell	29	54

7.4.9   Error handlers

Like most estimation algorithms, there may be occasions when the PVE algorithm may encounter processing error during computation and fail to produce accurate estimates. This may happen in the case of low SNR is presented (<10 dB) or when the cell size is small (radius < 200 m). The processing errors encountered are: cell size is small (radius < 200 m). The processing errors encountered are:

• Jacobian matrix becoming singular during iteration.

- Dividing by zero error when the MS position estimate during iteration is very close to a BS location (<2 m).

Several error handlers are introduced in the PVE algorithm to minimize the occurrence of these errors. The first error can be greatly reduced by constantly observing the Jacobian matrix using a reciprocal matrix condition estimator (RCOND) available in MATLAB. When the matrix is estimated as approaching singularity, the iteration is stopped and the latest result is used. The second error can be avoided by shifting the MS position estimate during iteration 5 m away from the BS location whenever the situation occurred and allows the iteration process to continue.

8 Smart antennas: mobile station antenna beamforming

Ng Kim Chong, Stetson Oh Kok Leong, P.R.P. Hoole and E. Gunawan

In this chapter we shall introduce the concept of smart or intelligent antennas in cellular communications. In a smart antenna the antenna beam is dynamically changed to enhance the system performance. In particular, by controlling the signal strengths at each element of an array antenna, by changing the weights of an adaptive antenna algorithm, the directivity of the antenna is dynamically controlled. We shall focus on the use of such an antenna on a mobile station, whilst remembering that the same principles may be applied for a base station smart antenna. Amongst the advantages of using smart antennas, the following are the most important:

- Increasing the channel capacity through frequency reuse within steerable beams. Since the power required is much less than a fixed antenna, resulting in a lower carrier-to-interference ratio, the smart antenna can allow channels to reuse frequency channels. Space division multiple access (SDMA) with smart antennas allows for multiple users in a cell to use the same frequency without interfering with each other, since the BS smart antenna beams are sliced to keep different users in separate beams at the same frequency.
- Increasing communication range without increasing battery power. The increase in range is due to a bigger antenna gain with smart antennas. This would also mean that fewer BSs may be used to cover a particular geographical area.
- Reducing multipath, cochannel interference and jamming signals by forming null points in the direction of unwanted signals. Hence the link quality can be improved. This could also enable the smart antenna beams to be always focused on the hot spots where the number of subscribers is large in a given area of a cell.
- Better tracking of the position and velocity of the mobile stations.

The position–velocity estimator (PVE) algorithm presented in Chapter 7 is further enhanced to include mobile station (MS) antenna beamforming. This is a crucial aspect of smart antennas in cellular communications. The MS estimates its own position and velocity, and simultaneously optimizes its antenna beam for reception and transmission. First, the possibility of combining the PVE algorithm and the LMS beamforming algorithm to perform beamforming and position–velocity estimation is investigated. Next and more importantly, an accurate single module beamforming with position–velocity estimator (BFPVE) algorithm is designed using the principle of maximum likelihood estimation. Based on a two-element antenna array, the proposed algorithms are tested in MATLAB^{TM} for different channel conditions.

8.1   MS adaptive beamforming

Most of the research effort on the use of adaptive antennas in mobile communications has been concentrated on the BS antennas. The objective of adaptive beamforming is to maximize the signal to interference and noise ratio (SINR) of the received desired signal, by maximizing the strength of the desired signal while reducing the adverse effects of interference sources. Although it is a practical, effective and hence a widely applied technique for reducing the effect of multipath fading, it is rarely used at a MS due to cost, size and available power of the MS. For hand-held units that are small lightweight pocket communicators, any additional hardware or software must be minimized while maximizing its functionality. Furthermore, it must remain affordable for widespread market acceptance. A single BS often serves hundreds to thousands of MSs. Therefore it is more economical to add equipment to base stations rather than at each MS. With the economy of scale on its side, it appears that BS complexity may be the plausible trade space for achieving the requirements of the next generation wireless system.

However, the possibilities of performing adaptive beamforming at the MS cannot be ignored. Besides, the algorithm proposed here deals with adaptive beamforming and position estimation as a whole. As the standard of living increases, subscribers expect a variety of services from their radio operators. Self-positioning at the MS can provide just that, with a variety of cellular services never thought possible. Future subscribers may find extra services such as location navigation and E-911 useful despite the additional cost they may incur. And with adaptive beamforming, it is reasonable to assume that the overall size of the hand-held unit is not increased significantly, if not reduced. Although an array antenna may occupy more space than the normal single antenna, this can be offset by the fact that less transmission power is required when beamforming is used, and less transmission power means fewer and smaller batteries required for the hand-held unit. With technology advance, size will not be a major problem in the near future. Further, in the system we proposed here, the number of antenna elements is kept to two. The signal processor is attached only to one of the two elements. This keeps hardware and software complexity to a minimum.

The BS antenna is always mounted on relatively high ground (e.g. on top of a building) and elevated above the local scatters. The MSs (either vehicular or hand-held) are below or at the same level as the local scatters. Thus MS antenna beamforming can contribute more significantly to reducing the interference from the local scatterers.

Of interest are also the potential health effects of radio frequency (RF) fields absorbed by the user's body as shown in Fig. 8.1(a). With a single antenna, the radiation pattern is omni-directional and the user's body may absorb more than half of the radiated power. Possible ways of reducing the damaging effects of microwave radiation are: the handset may be held some distance away from the ear (since power radiated is inversely proportional to the distance^2, i.e. $\propto 1/r^2$) or the use of the handset may be daily limited to about 5 minutes. A better method, which will improve safety and reduce required battery power, is the following: with antenna beamforming it is possible to steer the beam away from the human body as shown in Fig. 8.1(b).

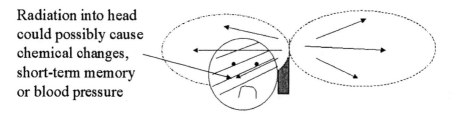

Radiation into head could possibly cause chemical changes, short-term memory or blood pressure

Single antenna - omnidirectional radiation

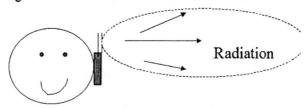

Radiation

Directional antenna - maximum radiation direction controlled

(a)

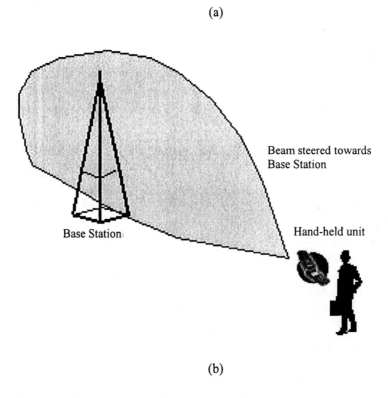

Beam steered towards
Base Station

Base Station

Hand-held unit

(b)

Figure 8.1: (a) Omni-directional and directional mobile phone radio frequency radiation. (b) The problem of dynamically directing the antenna beam to the base station.

However, the possibility of beamforming away from the human body may not always be feasible, as this will have to depend on the channel conditions, type of array antenna used, etc., but this may be used as one of the criteria when performing beamforming.

8.2 Array antenna

The adaptive beamforming system makes use of the antenna array to perform signal separation and interference rejection. Hence, to effectively execute these functions, we need to understand the characteristics of the antenna array and how the radiation pattern or beam can be controlled effectively. We shall ignore mutual coupling between elements; the effects of mutual coupling in an array antenna were discussed in Section 3.4. We shall also assume that the bandwidth (30 kHz) of the signal is small compared to the carrier frequency (e.g. 1980 MHz).

An antenna array consists of a set of antenna elements that are spatially distributed at known locations with reference to a common fixed point. By controlling the phase and amplitude of the exciting currents in each of the elements, it is able to electronically steer the main lobe or beam in any particular direction. The antenna elements can be arranged in different geometry, with linear, circular and planar arrays being very popular. In the case of a linear array, the antenna elements are aligned along a particular axis or straight line. If the spacing of the antenna is equally spaced, it is called a *uniformly spaced linear array*. In a circular array, the elements are arranged around the circumference of a circle with the center as the fixed reference point. In the case of a planar array, the antennas are distributed equally on a single plane.

The radiation pattern of an array is determined by the radiation pattern of its individual elements: their orientation and relative positions in space, and the amplitude and phase of the exciting current. If each element of the array is an isotropic point source, then the radiation pattern of the array will solely depend on the geometry and excitation current. This radiation pattern is called the array factor. If each of the elements of the array is similar but not isotropic, by the principles of pattern multiplication, the radiation pattern can be computed as a product of the array factor and the individual element pattern.

An array antenna formed by using two infinitesimal dipoles separated by distance d is considered and its resultant electromagnetic model is derived. Element 2 of the array antenna includes a weight component that provides the phase shift δ necessary for steering the beam of the antenna in any desired direction (see Fig. 8.2). An observation point $P(r,\theta,\phi)$ in the far-field region of the array antenna is defined in the Cartesian coordinate for the purpose of analysis.

The radiation field of an infinitesimal dipole antenna is defined in eqns (1) and (2) as

$$E(r,z,t) = \frac{\mu_0}{4\pi}\left[\frac{z_j - z_1}{\sqrt{r^2 + (z_j - z_1)^2}} - \frac{z_j - z_2}{\sqrt{r^2 + (z_j - z_2)^2}} \right]\frac{d[I]}{dt}, \tag{1}$$

$$[I] = \mathrm{Re}\left(I_0 e^{jw(t - R/c)} \right). \tag{2}$$

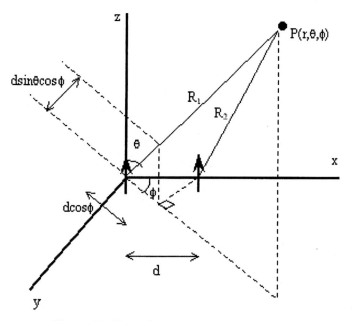

Figure 8.2: Two-element antenna array.

The resultant electric field intensity of the two elements is given by

$$E_T = E_1 + wE_2,$$
(3)

where

$$w = e^{j\delta}.$$
(4)

In the far zone, the magnitudes of E_1 and E_2 are assumed to be equal. This assumption is not valid for the phase of the two signals, since a small difference in distance resulted in large changes in phase.

$$E_T = \frac{\mu_0}{4\pi} \left[\frac{z_j - z_1}{\sqrt{r^2 + (z_j - z_1)^2}} - \frac{z_j - z_2}{\sqrt{r^2 + (z_j - z_2)^2}} \right] \frac{d[I]}{dt} \left[1 + e^{j(kd \sin\theta \cos\phi + \delta)} \right].$$
(5)

Using Euler's formula and letting $\psi = kd \sin\theta \cos\phi + \delta$, the resultant field can be expressed as

$$E_T = \frac{\mu_0}{4\pi} \left[\frac{z_j - z_1}{\sqrt{r^2 + (z_j - z_1)^2}} - \frac{z_j - z_2}{\sqrt{r^2 + (z_j - z_2)^2}} \right] \frac{d[I]}{dt} \left[e^{j\psi/2} \, 2\cos\left(\frac{\psi}{2}\right) \right].$$
(6)

Defining

$$AF = 2\cos\left(\frac{\psi}{2}\right),$$

(7)

eqn (6) becomes

$$E_T = E_1\, AF\, e^{j\psi/2}.$$

(8)

In a more general form where each element includes a weight component, the total output of an N-element array with complex weight is given by

$$E_T = \sum_{i=1}^{N} w_i E_i.$$

(9)

8.3   Basic concept of adaptive antenna

A two-element antenna array separated by half a wavelength is considered, and is shown in Fig. 8.3.

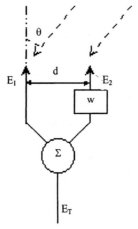

Figure 8.3: Basic concept of array antenna.

Assume that the desired signal arrives from an angle of θ_1 and the interference arrived at the array from an angle of θ_2. The radiation pattern can either be formed to maximize the signal reception or to eliminate the interference. To achieve both functions requires the radiation beam to be very narrow and also the angle of separation between the desired signal and interference to be large. From eqn (9), assuming no desired signal is present, the output of the array must be zero with the presence of interference,

$$\left| E_T \right| = 0.$$

(10)

To achieve the above requirement, the complex weight must be

$$w = -e^{-jkd \sin \theta_2 \cos \phi_2}.$$ (11)

Similarly, to achieve maximum output by considering the desired signal only, the resultant output from the array in the desired signal direction will be

$$|E_T| = 2.$$ (12)

Hence, the complex weight can be calculated as

$$w = e^{-jkd \sin \theta_1 \cos \phi_2}.$$ (13)

Both conditions require that the direction of the desired signal and interference must be known beforehand, which may not be a valid assumption in practice. Based on the knowledge of the characteristics of the transmitted signal, adaptive algorithms are developed to estimate the direction of arrival and automatically adjust the complex weight according to certain criteria and cost functions.

One of the most common algorithms, the least mean square (LMS) algorithm, proposed by Widrow and Stearns (1985) will be discussed in the following section.

8.4   Adaptive algorithm

A typical adaptive beamformer is shown in Fig. 8.4. In all practical communication channels or links, the incoming signal $x(t)$ usually is a mixture of desired signal, noise and interference from other users. Based on a set of predefined criteria and cost functions, the algorithm optimizes the response of the beamformer or complex weight, such that the output signal from the array contains minimum noise and maximum desired signal. The cost function is inversely proportional to the quality of the signal at the output of the array antenna, so that when the cost function is minimized the signal quality is maximized.

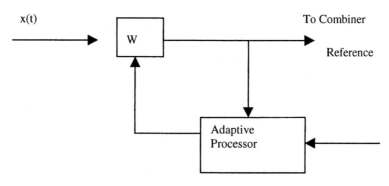

Figure 8.4: Typical adaptive beamformer block diagram.

The criteria can be classified into the following three:

- minimum mean square error (MMSE) criteria;
- maximum signal to interference ratio;
- minimum variance.

As the name implies, least mean square error criteria aims to minimize the mean square error between the array output and a reference signal. For the case of maximum signal to interference ratio, the complex weight is chosen such that the signal to interference ratio at the output will be maximized. For minimum variance criteria, the desired signal characteristics and its direction are usually unknown. Hence, the output noise variance is suppressed so that there is good reception of the desired signal.

All the above three criteria will resolve to the optimal Wiener–Hopf solution, which yields the same signal to interference ratio (SIR), but represented in different forms. The following section briefly describes the least mean square criteria, which leads to the least mean square (LMS) algorithm.

8.4.1   Minimum mean square error criteria

Consider a uniformly spaced linear array as shown in Fig. 8.2. Assume that a single desired signal exists and the communication link is polluted only by a single interference source. The desired signal $s(t)$ arrives at the array with a spatial angle θ_d, ϕ_d and the interference $i(t)$ reaches the array at an angle of θ_i, ϕ_i. The array output $x(t)$ shown in Fig. 8.4 is

$$x(t) = s(t)v_d + i(t)v_i,$$
(14)

where v_d, v_i are the array propagation vectors represented by

$$v_d^T = \begin{bmatrix} 1 & e^{-jkd \sin \theta_i \cos \phi_d} \end{bmatrix},$$
(15)

$$v_i^T = \begin{bmatrix} 1 & e^{-jkd \sin \theta_i \cos \phi_i} \end{bmatrix}.$$
(16)

Including the complex weight w_i, the array output will be

$$y(t) = \sum_{i=1}^{N} w_i x_i.$$
(17)

In matrix form

$$y(t) = w^T x.$$
(18)

In least mean square criteria, the output is corrected by the complex weight such that minimum mean square error between the output and desired signal is achieved. Therefore, the transmitted signal must be known beforehand, which is an ambiguous statement. However, for many applications, the characteristics of the desired signal may be known with sufficient details. With this knowledge, the receiver can self-generate a reference signal which closely resembles the desired signal, or is highly correlated with the original signal. The criteria can be represented as

$$\varepsilon^2(t) = [r(t) - w^T x(t)]^2. \tag{19}$$

Taking the expectation on both sides and rearranging the equation,

$$E[\varepsilon^2(t)] = E[r^2(t)] - 2w^T S + w^T R w, \tag{20}$$
$$S = E[x(t)r(t)], \tag{21}$$
$$R = E[x(t)x^T(t)], \tag{22}$$

where R is usually referred to as the covariance matrix. The minimum mean square error is given by setting the gradient vector of eqn (20) to zero,

$$\frac{d\{E[\varepsilon^2(t)]\}}{dw} = 0, \tag{23}$$

$$\frac{d\{E[\varepsilon^2(t)]\}}{dw} = -2S + 2Rw. \tag{24}$$

Hence, the optimal weight that can achieve minimum mean square error is

$$W_{opt} = R^{-1}S. \tag{25}$$

We note that in the MMSE algorithm the solution given in eqn (25) gives us a single weight vector that is optimal over the collection of measurements over a certain time window. The solution depicted in eqn (25) is also referred to as the Wiener–Hopf solution or the optimal Wiener solution. To obtain this solution, the covariance matrix R and correlation matrix S will be required. In practice, both quantities are difficult to compute or may even be impossible to get due to lack of information.

8.4.2   Least mean square algorithm

The mean square error criteria shown in the previous section only served as the cost function in the adaptive algorithm. The various ways of implementing the adaptive algorithms, which select the optimal weight for the beamformer, will have a much greater impact on the system performance. Mainly speed of convergence, computational power and time required, and hardware complexity are the most critical factors in consideration for the type of algorithm used.

The most common adaptive algorithm being studied is the LMS algorithm. With a reference signal available, the LMS algorithm is simple in nature and requires the least amount of computation time. In the MMSE solution, the error was minimized for a collection of measurements averaged over a period of time. In the LMS approach, the output of the antenna array and a desired response over a finite number of samples is minimized.

To recursively compute and update the weight vectors, the LMS algorithm utilizes the method of steepest descent. According to this method, the updated value of the weight vector at time $n + 1$ is computed using the following relationship:

$$w(n+1) = w(n) + \tfrac{1}{2}\mu[-\nabla(E\{\varepsilon^2(n)\})]. \qquad (26)$$

The error surface $E\{\varepsilon^2(t)\}$ is always viewed as a bowl or rather as a quadratic function. The definition of such a function is important, as it implies that in the error function only a single minimum point exists that will give the optimal solution. Intuitively, one concludes that if the direction of the gradient movement is negative, it will be heading towards the bottom of the bowl, which is the minimum point of the error function.

Recall from eqn (24) that, in order to compute the relationship in eqn (26), one requires a knowledge of the covariance matrix R and the correlation matrix S, which is not always available.

Thus, the most obvious strategy is to use the instantaneous gradient estimation for the computation in eqn (26):

$$\frac{d[\varepsilon^2(n)]}{dw} = -2\varepsilon(n)x(n). \qquad (27)$$

Substitute into eqn (26),

$$w(n+1) = w(n) + u\varepsilon(n)x(n). \qquad (28)$$

This is the LMS updating algorithm presented by Widrow and Stearns (1985). Even under non-stationary signal condition, the algorithm could converge to the optimal Wiener solution, at the expense of slow convergence rate.

The LMS adaptive algorithm is simple and easy to implement. However, it requires a reference signal in order to perform computation. Normally, a pilot signal or training sequence will be transmitted over the channel in order to pre-train the weight. In this way, the performance of the communication system will be degraded. Further, under noisy signal environment like with mobile communication, the algorithm tends to converge of a much slower rate, typically of 200-iteration loop. This is unacceptable for real time application like voice channel communication. One drawback in the MMSE and LMS methods is that the desired output must either be known or estimated. This knowledge is often obtained by sending a known training sequence on a periodic basis, where the sequence is known to both the transmitter and the receiver. Since this training sequence carries no useful data except for training the adaptive receiver, the frequency spectral it occupies is wasted. Alternatives to MMSE and LMS that do not require such training sequences are the decision-directed algorithms or the blind adaptive algorithm, where an attempt to restore some known property of the received signal is made.

8.5   Electromagnetic model

As derived in Section 8.2, the E-field of a two-element array is simply the multiplication of a single element pattern with an array factor. From eqn (6),

$$E_{\mathrm{T}} = \frac{\mu_0}{4\pi} \left[\frac{z_j - z_1}{\sqrt{r^2 + (z_j - z_1)^2}} - \frac{z_j - z_2}{\sqrt{r^2 + (z_j - z_2)^2}} \right] \frac{\mathrm{d}[I]}{\mathrm{d}t} \left[e^{j\psi/2} \, 2\cos\left(\frac{\psi}{2}\right) \right]. \qquad (29)$$

The beamforming signal processor operates only on the magnitude of the measured electric field strength. Hence eqn (29) is reduced to

$$|E_{\mathrm{T}}| = A_0 \left[\frac{z_j - z_1}{\sqrt{r^2 + (z_j - z_1)^2}} - \frac{z_j - z_2}{\sqrt{r^2 + (z_j - z_2)^2}} \right] \left| 2\cos\left(\frac{\psi}{2}\right) \right|, \qquad (30)$$

where

$$A_0 = \frac{\mu_0 \omega I_0}{4\pi} = 10^{-7} \omega I_0 ,$$

$$\psi = kd \sin\theta \cos\phi + \delta,$$

$$I_0 = \sqrt{\frac{P_{\mathrm{r}}}{R_{\mathrm{r}}}} ,$$

$|2\cos(\psi/2)|$ is the magnitude of the array factor, P_{r} is the power radiated and R_{r} is the radiation resistance of the transmitter antenna. All parameters are defined in Figs 8.4 and 7.2.

8.6   Beamforming with position and velocity estimator (BFPVE)

A single module that performs beamforming and position–velocity estimation simultaneously in each iteration process is being proposed as shown in Fig. 8.5.

Figure 8.5: Block diagram of BFPVE.

The measured instantaneous electric field strengths E_{m} in the presence of additive white Gaussian (AWG) noise is modeled by

$$E_{\mathrm{m}} = E + n, \qquad (31)$$

where E is a function of r and w is defined in eqns (1) and (30).

Firstly, an initial estimated r and w are assumed to be obtainable (described in Section 8.10). From this assumption, E can be made a function of only one of the two parameters (r or w). Thus an estimator for r and w can be derived similar to the deviation of the PVE algorithm.

Through series expansion, E can be expressed as

$$E(a) = E(a_0) + E'(a_0)\,\Delta a + 0.5\,E''(a_0)\,\Delta a^2 + \cdots, \tag{32}$$

where $a = r$ or w. Ignoring higher order terms and writing $E_0 = E(a_0)$, let

$$J_0 = E'(a_0) = \frac{\delta E_0}{\delta a_0}, \tag{33}$$

$$E(a) = E_0 + J_0\,\Delta a. \tag{34}$$

Therefore the noise term is given by

$$n = (E_m - E_0) - J_0\,\Delta a = \Delta z - J_0\,\Delta a. \tag{35}$$

In order to minimize $U(\Delta a) = n^T n$, let $\partial U(\Delta a)/\partial a = 0$, which with eqns (33) and (34) yields

$$\Delta a = (J_0^T J_0)^{-1} J_0^T \Delta z. \tag{36}$$

Therefore, a can be estimated using a maximum likelihood estimator of the following form:

$$a^{n+1} = a^n + (J_n^T J_n)^{-1} J_n^T (E_m - E_n). \tag{37}$$

Replacing a by parameters r and w, eqn (37) can be written as

$$r^{n+1} = r^n + (Jr_n^T Jr_n)^{-1} Jr_n^T (E_m - E_n), \tag{38}$$

$$w^{n+1} = w^n + (Jw_n^T Jw_n)^{-1} Jw_n^T (E_m - E_n), \tag{39}$$

where E_m is a column vector that represents the measured electric field strengths at the BSs. Jr_n and Jw_n are the Jacobian matrix of the electric field strengths E_n with elements given by

$$Jr = \frac{\partial E}{\partial r} = A_0 r \left\{ \frac{Z_2 - Z_j}{\left(r^2 + (Z_j - Z_2)^2 \right)^{3/2}} - \frac{Z_1 - Z_j}{\left(r^2 + (Z_j - Z_1)^2 \right)^{3/2}} \right\} \left| 2\cos\frac{\psi}{2} \right|, \tag{40}$$

$$Jw = \frac{\partial E}{\partial w} = E_2. \tag{41}$$

The algorithm defined by eqns (38) and (39) is named BFPVE I. Both equations undergo an iteration process simultaneously until the required convergence accuracy is achieved for both r and w.

An alternative approach to generate the required weight is to represent the weight w as a function of the parameter r. Both R and θ shown in Fig. 7.2 and eqn (30) can be represented as a function of parameter r:

$$R^n = \sqrt{\left(z_2 - z_j\right)^2 + \left(r^n\right)^2} \, , \tag{42}$$

$$\theta^n = \sin^{-1}\left(\frac{r^n}{R^n}\right), \tag{43}$$

where the superscript n represents the value at each iteration cycle number n.

To maximize the desired signal strength, the w required is just the complex conjugate of the element factor of antenna. Hence

$$w^n = \left(e^{jkd \, \sin \theta^n \, \cos \phi^n}\right)^* , \tag{44}$$

where * represents complex conjugate. The algorithm defined by eqns (38) and (44) is named BFPVE II. Hence, BFPVE I uses separated iteration equations for r and w whereas BFPVE II uses only a single r estimator by setting w as a function of r.

8.7   Simulation scenario

In order to evaluate the performance of the algorithms, simulations are performed in MATLAB™. A wireless communication scenario is shown in Fig. 8.6.

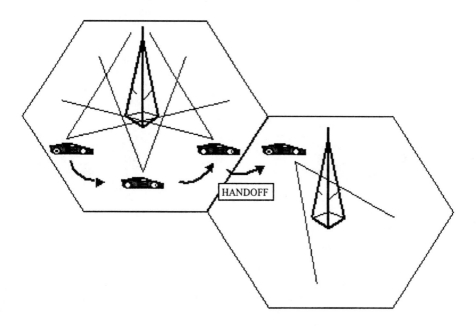

Figure 8.6: MS antenna beamforming.

The simulation scenario used here is basically the same as in Chapter 7 used to evaluate the performance of the PVE algorithm, with some additional factors considered for MS adaptive beamforming. As the MS travels along the route, it always beamforms towards the base station transceiver (BTS) that handles the wireless communication link between the MS and the radio network. The MS beam pattern constantly steers towards the direction of the BTS so as to maximize the reception of the desired signal while minimizing interference signals from other directions. As it moves into another cell, handoff is performed and the beam is steered towards the new BTS as shown in Fig. 8.6.

Reception from three BSs with the strongest field strength is used to compute the parameter r. Thus r is a vector with three elements $[r_1, r_2, r_3]$, each value is the horizontal distance between a BS and the MS. As the weight w is derived for maximizing the reception of the desired signal, its value is only dependent on the field strength from that particular BTS where the desired signal is transmitted. Thus w is a vector with three elements of the same value.

8.8   Channel models

The channel models used previously in Chapter 7 are based on the assumption of Gaussian distribution noise. However, when beamforming is performed, that assumption is no longer valid. The beam pattern limits the directions of reception of interference signals. To adjust to this condition, the noise may be multiplied by the array factor that forms the beam pattern (see Fig. 8.7).

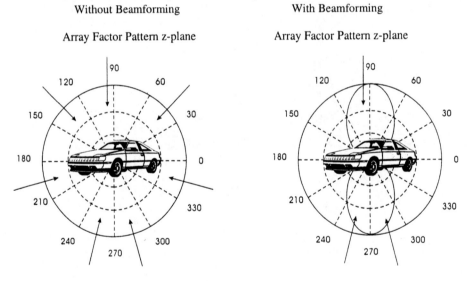

Figure 8.7: Noise channel.

The measured electric field is given by

$$E_m = E + \text{AF } N_{\text{AWG}}(m, \sigma), \qquad (45)$$

where E is the E-field model eqn (40), AF is the array factor of the MS antenna, N_{AWG} is the additive Gaussian distributed noise with mean m and standard deviation σ.

8.9   Antenna radiation pattern

The formation of the beam pattern depends on the array factor AF that is given by

$$ \text{AF} = 2\cos\left(\frac{kd\sin\theta\cos\phi+\delta}{2}\right), \tag{46} $$

where k is the wave number $2\pi/\lambda$ where λ is the wavelength, d is the distance spacing between the two elements, θ is the elevation angle of the BS to MS and is a function of parameter r, ϕ is the direction of the desired signal from the BTS and δ is the phase difference between the two elements contributed by the weight.

Both k and d are fixed parameters while θ, ϕ and δ are variables depending on the position of the MS. During iteration, the weight converges to a complex value that is used to compute the phase δ. The angle θ is a function of parameter r as indicated in eqn (43). To simplify computation, angle ϕ is assumed to be a known value. However, it can be shown that ϕ can be estimated during each iteration cycle by the iterating r values, since the calculated r values represent the position estimation that is converging closer towards the actual MS position. Hence, with the location of the BTS known, the direction of arrival of the desired signal ϕ can be estimated.

The overall radiation pattern is just the multiplication of the single element pattern by the array factor pattern. The single element pattern is different in the H-plane and E-plane. The single element radiation pattern in the H-plane is just a circle, whereas it has regions of zero radiation along the axis on which the dipole is placed. As the array factor pattern is identical for both the H-plane and E-plane, the overall radiation pattern in the H-plane is just the array factor pattern. The pattern in the H-plane, which is the xy-plane, will show clearly the steering of the beam pattern towards the BTS as the MS travels.

The overall antenna radiation pattern will have different shapes in the horizontal (H-) and vertical (E-) planes. If the maximum gain in the horizontal plane is G_H and the maximum gain in the vertical plane is G_E, then the maximum gain of the entire three-dimensional beam is given by $G_B = G_H G_E$. In general, the carrier to interference and noise ratio (CINR) is proportional to the horizontal beam gain G_H. The CINR is typically given by CINR = $(QSG_H)/(aN)$, where Q is the reuse factor, S is the spreading factor, a is the vocoder rate and N is the number of users. The reuse factor Q (<1) is the ratio of received power from all users within a cell to the total interference from all users. The spreading rate associated with CDMA systems is given by S = chip rate/information symbol rate. For a pseudo-noise (PN) chip rate of 1.2288 Mchips/s, and repeated code symbol rate (RCSR) of a convolutional encoder of 28.8 ksps, the spreading rate is about 43. The vocoder rate a is associated with the reduction digital signal output rate when a speaker in a voice communication systems is silent. The periods over which the MS does not transmit is given by a (typically 0.45), resulting in a reduction of the multiple access interface level at the BS. For a CINR of 10 dB required for each user, for a smart antenna at the BS with $G_H = 5$ dB the system can support about 40 users, whereas if an omni-directional BS antenna was used

($G_H = 0$ dB) only about 15 users could have been supported by the BS antenna. If G_H of the BS antennas was increased to 10 dB, then more than 100 users could have been supported by the BS receiver. Further discussion of this subject from a BS antenna perspective is found in Liberti and Rappaport (1999).

8.10   Initial values

The procedure for initial r estimation is the same as described in Chapter 7. The initial weight (w at $n = 0$) estimation can be obtained by using the complex conjugate of the element factor which is a function of angles θ and ϕ.

Using antenna element 1 as reference (i.e. element factor = 1), the total E-field measured is

$$E_T = E_m + we^{jkd \sin \theta \cos \phi} E_m, \tag{47}$$

where $e^{jkd \sin \theta \cos \phi}$ is the element factor of element 2.

At the start of iteration, initial weight estimation can be obtained by using the complex conjugate of the element factor.

8.11   Simulation results

The iteration begins with the initial values of r and w, and proceeds until a convergence accuracy of 10^{-4} is achieved. BFPVE I needs both r and w estimators to achieve the required criteria before the iteration loop is stopped, whereas BFPVE II depends only on the r estimator.

Figures 8.8 and 8.9 show convergence characteristics of r and phase angle w for the AWGN channel at an SNR of 10 dB. The average number of iteration loops for convergence is 6. The accuracies of the estimated r vector [r_1, r_2, r_3] are compared with those of the actual values in Table 8.1 for MS positions 1, 8 and 15.

Table 8.1: Comparison of estimated r with actual r.

MS position	Estimated r vector (km)	Actual r vector (km)
Position 1	[0.808, 1.028, 1.161]	[0.849, 1.059, 1.097]
Position 8	[0.613, 1.139, 1.465]	[0.623, 1.102, 1.457]
Position 15	[0.873, 0.997, 1.281]	[0.840, 0.955, 1.245]

The result shows that the estimated r is reasonably accurate when compared to its actual values. This result is similar for both BFPVE I and BFPVE II. From Figs 8.10–8.12 and Table 8.2, the position and velocity estimates obtained using the two estimators can be compared. The results show that the accuracy of BFPVE I is similar as compared to BFPVE II. For example for AWGN with an SNR of 10 dB, BFPVE I has an average position estimation error of 55 m as compared to approximately 56 m for BFPVE II. From Table 8.2, it can be seen that under a dominant reflected path channel, the estimation is less accurate for all cases. However, using suitable filtering techniques, the accuracy can be significantly improved.

Table 8.2: Simulation results.

Channel model (SNR = 10 dB)		Average position error (m)	Average velocity error (km/h)
AWGN	BFPVE I	55	20
	BFPVE II	56	24
Rayleigh	BFPVE I	70	26
	BFPVE II	74	27
Reflected path	BFPVE I	147	43
	BFPVE II	143	42
Rician	BFPVE I	85	33
	BFPVE II	80	28

Figure 8.11 shows the H-plane radiation pattern for the estimated MS positions. It clearly shows how the beam pattern is steered as the MS travels along the route marked by the dotted line. The beam pattern consists of two lobes, with one of the lobes covering the direction of the desired signal, as can be seen more clearly in Fig. 8.12 which shows the three-dimensional radiation pattern. Besides the additional smaller lobe, the main lobe beam width is quite wide, covering more than the intended direction. This is due to the fact that only two elements are used for the array. For a narrower and better-focused beam, three or more elements are required. With more elements, it is possible that other criteria such as nulling of interference signals from specific directions can also be satisfied. However, for MS beamforming with position estimation, a reasonable wide beam covering at least three BSs is necessary. If the beam is too narrow, the signal from other BSs may be too weak to be used for computation. Furthermore, increasing the number of elements makes the MS antenna cumbersome.

Hence, from the results shown, the use of an array antenna at the MS allows its beam pattern to be adjusted such that the gain in the direction of the desired signal is maintained constant while having nulls in other directions. Thus interference signals or multipath signals with large delay spreads coming from these directions will be rejected. This helps to reduce co-channel interference (CCI) and inter-symbol interference (ISI) and improves the network capacity as the effective SINR is significantly increased.

MS position 1

Estimated r (km):
0.808, 1.028, 1.161

Actual r (km):
0.849, 10.59, 1.097

MS position 8

Estimated r (km):
0.613, 1.139, 1.465

Actual r (km):
0.623, 1.102, 1.457

MS position 15

Estimated r (km):
0.873, 0.997, 1.281

Actual r (km):
0.840, 0.955, 1.245

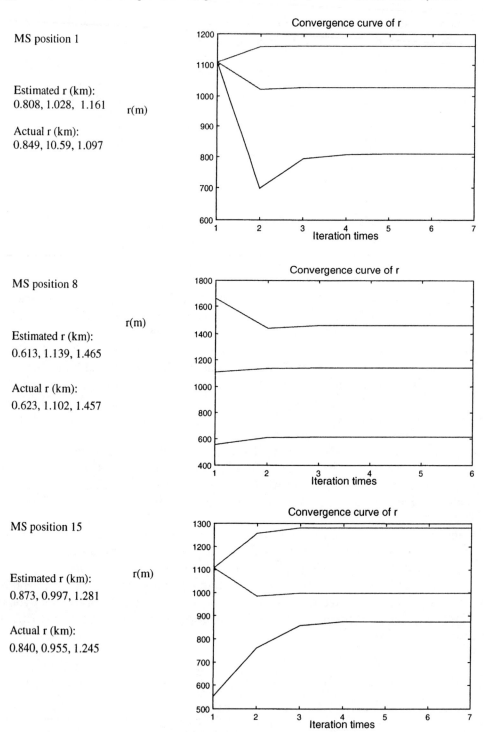

Figure 8.8: Convergence of parameter *r*.

MS position 1

MS position 8

MS position 15

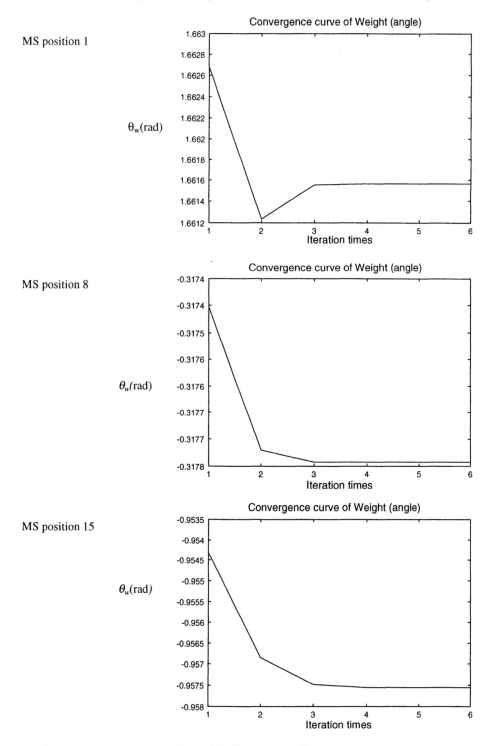

Figure 8.9: Convergence of parameter w.

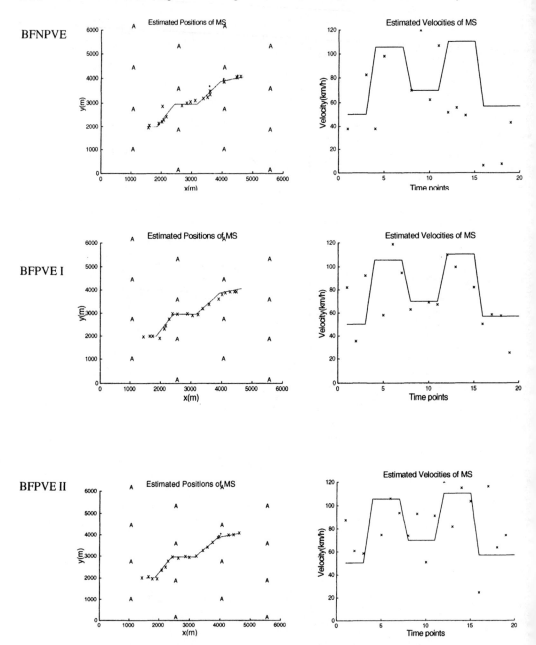

Figure 8.10: Simulation results (Rayleigh channel; SNR=10 dB).

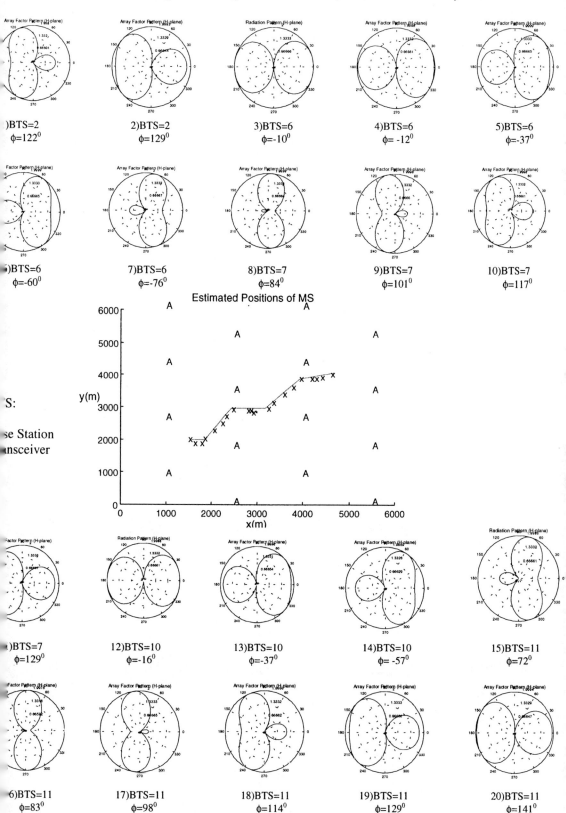

Figure 8.11: H-plane radiation patterns at different estimated MS positions.

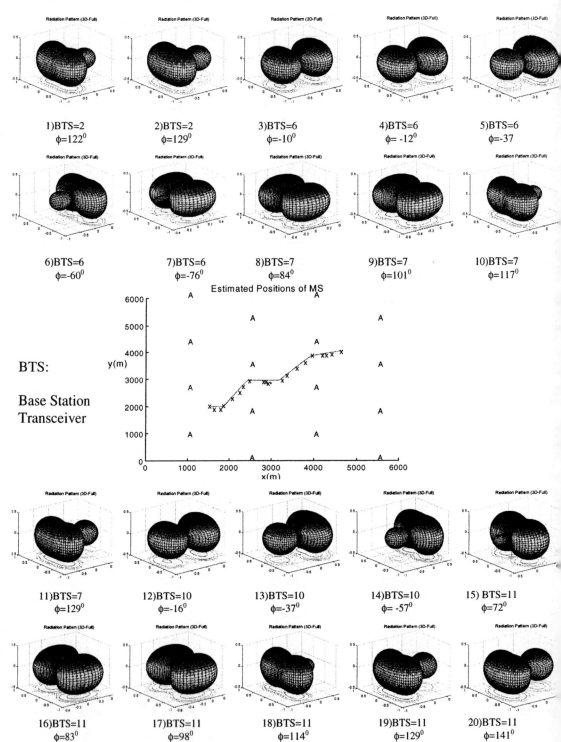

Figure 8.12: Three-dimensional radiation patterns at different estimated MS positions.

Transmission power required at the MS to perform the same basic communication functions is largely reduced. This can be easily seen by looking at the different coverage area of the beam pattern maintained by a single dipole and a two-element array antenna as shown in Fig. 8.13. Less power is wasted in covering regions besides the region of interest. The significant reduction in power required helps to extend battery life as well as provides possible reduction in the size and cost of the batteries required at the MS.

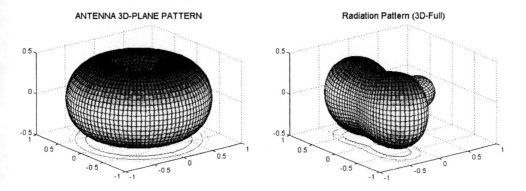

A single dipole antenna   Two-element array antenna

Figure 8.13: Advantage of array antenna: power reduction.

9 Smart antennas: base station antenna beamforming

Ng Yin Yoon, Edmund Yap Aik Boon and P.R.P. Hoole

9.1 Introduction

9.1.1 Smart antennas

In cellular systems such as GSM or IS-95, antenna arrays can be used at base stations (BSs) to increase user capacity and provide other advantages such as improving voice quality, extending battery life in handsets and increasing data transmission rates. Array antennas have been employed in military systems for many years as a counter measure for deliberate, high power jammers. In Chapter 8 we considered the use of smart antennas in the mobile station (MS). In this chapter we shall consider its use in the base station. When smart antennas are used in MSs, there is a constraint on the number and size of the antenna we could use since the MS should be as small and as light as possible. Further, the MS smart antenna needs to be communicating with only one BS at a time. The motivation for using smart antennas in the MS is to increase the range over which the MS antenna may communicate, to reduce interference with other users and to reduce the power requirement on the batteries used in the MS. When smart antennas are used in BSs, we may use a large number of elements to form the array antenna. However, one stringent requirement is that a single BS smart antenna may need to communicate with more than one user at the same time. Indeed one reason for using smart antennas in BSs is to increase the number of users the BS antenna may service simultaneously, that is to increase the capacity of the system. In this chapter we shall consider the use of BS smart antennas to improve on the range and capacity of the cellular system.

An array antenna is made up of N antenna elements that are spatially arranged at specified locations about a common fixed point. Using the principle of superposition, the radiation pattern of an array antenna is the summation of the radiation pattern of its antenna elements. The beams may be formed and directed by changing the electronic current phase, current amplitude or both current phase and amplitude of the currents at each antenna element. Control of current phase and amplitude is performed to maximize gain in a specific direction, to minimize sidelobe and to increase the range of the beam. In mobile communications, the ability to electronically control the antenna beam may be exploited to improve the antenna and communication system performance.

The spatial arrangement of the antenna elements in an array antenna has received less attention in recent times. Obviously the geometry of the array antenna will have an impact on the beam generated. The linear, circular and planar arrays are the three most commonly used geometries in array antennas, but it has been found that the geometry of the array does not have a significant effect on the efficiency of the antenna beam. It is the beamforming algorithm, which may be implemented on software or hardware, that largely determines the efficiency of the array antenna.

Beamforming of array antennas allows for achieving diversity in wireless communications without having to resort to more than one location for antennas. Antennas located in different locations in a neighborhood have been used to increase diversity and thus to combat fading in microwave mobile radio systems.

The following are two common forms of array antennas:

1. *Switched beam system.* An electronic switch that switches out or, in some of the antenna elements in an array antenna, is used in switch beam systems (SBS) to select the best beam for a given user scenario. Although it is not easy to null any interfering signal that is close to the desired user signal with SBS, it is a much cheaper and simpler system than the smart antenna system.
2. *Adaptive antenna system.* An adaptive antenna system is able to achieve better performance. The weight in an adaptive array system (AAS) may be adjusted to point the tip of the beam towards the desired signal, and interferers are nulled by adjusting electronic weights that control the amplitude or phase of the currents at the antenna elements. It may not be as fast as the SBS, but it provides a more optimized beam for a given situation of users and interferers.

For convenience the array antennas are subdivided into two classes. The first is the phase array antennas where only the phase of the currents is changed by the weights. The second class of array antennas is the smart antennas where both the amplitude and phase of the currents are changed to produce a desired beam. One major disadvantage with the phase array systems is that the optimum weights to maximize the signal to noise ratio (SNR) are not computed due to the limit on the degree of freedom. However, it is a much faster system than the smart antennas.

Smart antenna gain is much larger than the static array antenna since multipath effects may be reduced. Further improvement on the transmission or reception gain may be achieved by increasing the number of antenna elements in the array antenna. Such an increase in antenna elements is possible in the base station thus permitting an increase in the tolerable path loss. The ability of the smart antennas to provide separate beams for different users will result in an improvement in the system capacity. The increasing demand for communication is expected to exceed the projected demand with the introduction of the personal communications network (PCN). Reduction in the co-channel interference from other users within its own and neighboring cells will lead to an increase in the BS system capacity. Thus the smart antenna increases the system capacity by reducing interference at the BS.

The combination of the multiple beams that an array antenna can produce and the possibility of dynamically controlling the beam may be further exploited to improve system performance. Smart antennas can also be used to achieve a fourth form of multiple access: spatial division multiple access (SDMA). In the SDMA system, users that may be clustered in different areas of a cell may be spatially lumped into clusters. Each cluster may now be assigned a different frequency/time slot. Thus SDMA allows multiple users in the same cell to use the same frequency/time slot provided. Therefore by using smart antennas in SDMA to more effectively use the spectral resources, we may further improve on the system capacity.

For large array size for adaptive beamforming, allowing for full adaptivity may mean heavy computational burden and time to achieve convergence of all the weights. To minimize computational burdens in large array antennas, various decomposition methods such as singular value decomposition have been proposed. The use of partial adaptivity can reduce the cost and the heavy computational complexity required by a fully adaptive beamformer since only a part of the available degree of freedom is adapted. The speed of a partially adaptive beamformer can be increased because the number of the adaptive weights used is reduced.

Where the amount of interference is expected to be high, a smart antenna may need a canceler to help it to increase SNR. A smart antenna increases its directivity by nulling the interferer signals. It does this by automatically determining the direction of the desired signals and that of the interferer signals. If the number of interferers increases, however, then the smart antenna will be at a loss to cancel or null the interfering signals. The problem is one of obtaining a reliable reference signal to zoom in on the desired signal. A canceler added to the array antenna helps to generate a good reference signal even in the presence of heavy interference. The use of a canceler results in stable demodulation and it improves the error rate of decoded data even when there is heavy interference. As adaptive beamforming is sensitive to slight errors in array characteristics, robust algorithms that deal with quadratic equality together with linear equality have been investigated. The quadratic part of the equality will ensure the beamformer is not highly sensitive to small changes in amplitude, phase or position errors.

Smart antenna systems manage to mitigate the effect of imperfect power control in the CDMA system. In the CDMA system, it is crucial to ensure that all received signals are at approximately the same power level so as to reduce multiple access interference (MAI). Smart antenna systems help to isolate the signal of interest (SOI) from the different users, reducing the power control requirement. The implications of using smart antennas in CDMA wireless systems as far as power control is concerned have yet to be worked out more carefully. Smart antenna systems can have a very significant impact on improving the efficiency of CDMA systems. The CDMA system suffers greater interference (multipath fading) due to the usage of same frequency for all users. Furthermore, the CDMA system performance is affected by the near–far problem. In both these areas, of near–far effect and broadband interference, the smart antenna can help to greatly improve the performance of a CDMA system.

9.1.2   Handover in beamforming scenario

Handover of a mobile station (e.g. a mobile phone), also known as handoff, is the process of automatically transferring a call in progress from one cell to a different cell. Handover is performed to maintain continuity of a call without any cut or drop, taking into account the adverse effects of user movements, such as movement from one cell to another, drop in received signal strength below the minimum level required by the receiver or exceeding the timing advance adjustment capabilities.

Many different approaches to determine the criteria for handover have been considered. The parameters considered as indicators for a handover decision are bit error rate (BER), BS to MS distance, SNR, signal strengths and a combination of these. Among them, signal strengths based handoff algorithms are very popular, due to their simplicity and good performance. In a GSM system, signal strengths (time averages of received signal strength) at the six nearest base stations are compared and the control base station (CBS) is chosen with the strongest signal strength. With the movement of the MS, if the signal strength of the CBS falls below that of any of the other nearby five BSs by a threshold (also known as hysteresis level, e.g. 8 dB), a handoff will take place.

However, algorithms based on signal strength become more complicated in adaptive array antenna (AAA) systems, since the signal strengths will be strongly determined by the beam formed. This is not the case with the omni-directional or sectored fixed-beam antennas that were used for the first two generations of wireless systems. A new and simpler method of handoff for smart antennas will be proposed in this chapter.

9.1.3   Position–velocity estimator

There are several types of position estimators that have been developed and used in recent wireless transmission systems. These include the BS-based positioning, MS-based positioning and global positioning satellite (GPS). Whereas in Chapter 8 we considered MS-based positioning of the MS, in this chapter we consider BS-based positioning of the MS. MS positioning is important to enhance system capacity, for beamforming and to help in monitoring the call management between cells. The general techniques used for MS positioning make use of the signal strengths, the angles-of-arrival (AOA), the time-of-arrival or the combination of them, measured at several base stations.

With the use of smart antenna systems, the conventional MS position estimator has to be modified since it involves the array factor of the antenna radiation pattern. The array factor introduces more complications into MS position–velocity calculation. A unique position–velocity estimator for smart antennas, using maximum likelihood method, is described later in this chapter.

9.2   Smart antenna theory

9.2.1   Adaptive array antennas

Adaptive array antennas are used in mobile, wireless communications due to their ability to separate a desired signal from interfering signals. They are able to optimize the antenna array pattern by adjusting the elemental control weights until a prescribed objective function is optimized. Figure 9.1 shows a typical beamforming system. The choice of the weight vector w is based on the statistics of the signal vector $X(t)$ received at the array.

Consider an array antenna formed by using N infinitesimal dipoles as shown in Fig. 9.2. The dipole antenna elements are separated by distance $d = \lambda/2$. The desired signal $S(t)$ arrives from the angle θ_d, and the interference signal $I(t)$ arrives from the angle θ_i. Both signals $S(t)$ and $I(t)$ have the same frequency f. The signal from each element is multiplied by a variable complex weight, and the weighted signals are then summed to form the array output.

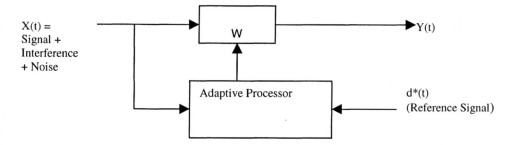

Figure 9.1: A typical beamforming system.

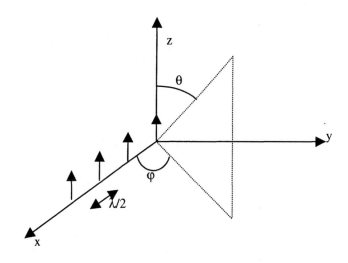

Figure 9.2: Antenna array arrangement.

The antenna array output is given by

$$X(t) = S(t)v + I(t)u \qquad (1)$$
$$= s + I, \qquad (2)$$

where v = the array propagation vector for the desired signal, and

$$v^{\mathrm{T}} = [1, e^{jkd\,\sin\theta d\,\cos\varphi d}, \ldots, e^{j(N-1)d\,\sin\theta d\,\cos\,\varphi d}], \qquad (3)$$

u = the array propagation vector for the interfering signal, and

$$u^T = [1, e^{jkd \sin \theta i \cos \varphi i}, \ldots , e^{j(N-1)d \sin \theta i \cos \varphi i}]. \tag{4}$$

If the desired signal $S(t)$ is known, one may choose to minimize the error or mismatch between the smart antenna output $w^H X(t)$, and the desired signal. The vector w is the weight array of the smart antenna. However for many applications, the characteristics of the desired signal $S(t)$ are known with sufficient detail to generate a signal $d*(t)$ which closely represents $S(t)$, or at least correlates with the desired signal to a certain extent. This signal $d(t)$ is known as the reference signal. The weights are then chosen to minimize the mean square error (MSE) between the smart antenna output and the reference signal.

$$\varepsilon^2(t) = [d*(t) - w^H X(t)]^2. \tag{5}$$

Taking the expected values of both sides of the above equation and carrying out some basic algebraic manipulation, we have

$$E\{\varepsilon^2(t)\} = E\{d^2(t)\} - 2w^H r + w^H R w, \tag{6}$$

where $r = E\{d*(t)X(t)\}$ and $R = E\{X(t)X^H(t)\}$. R is usually referred to as the covariance matrix. The minimum MSE is given by setting the gradient vector of $E\{\varepsilon^2(t)\}$ with respect to w equal to zero. This yields (see Chapter 4)

$$\nabla_w(E\{\varepsilon^2(t)\}) = -2r + 2Rw \tag{7}$$
$$= 0.$$

Therefore the optimum weight vector is given by

$$W_{opt} = R^{-1}r. \tag{8}$$

The above MSE criteria to obtain the weights are not the only optimization techniques that may be used. There are other criteria such as maximum SIR (signal to interference ratio) and minimum variance. All these three techniques are related to each other. This means that the choice of a particular criterion will not significantly affect the performance of the array antenna.

9.2.1.1   Least mean square algorithm

The most commonly used adaptive algorithm for continuous adaptation of the weights, and hence the beam, as the MSs move is the least mean square (LMS) algorithm. It is based on the steepest descent method, a well-known optimization method that recursively computes and updates the weight vector. According to the method of steepest descent, the updated value of the weight vector at iteration number $n+1$ is obtained by using the simple recursive relation

$$w(n+1) = w(n) + \tfrac{1}{2}\mu[-\nabla(E\{\varepsilon^2(n)\})]. \tag{9}$$

Using the MSE criteria,

$$w(n+1) = w(n) + \tfrac{1}{2}\mu[r - Rw(n)]. \tag{10}$$

However, an exact calculation of the gradient vector is not possible, since this would depend on a prior knowledge of both R and r. Thus it is preferable to use their instantaneous values obtained from

$$\overline{R}(n) = X(n)X^H(n) \tag{11}$$

and

$$\overline{r}(n) = d^*(n)X(n). \tag{12}$$

Hence the smart antenna weights can then be updated as follows:

$$\begin{aligned}\overline{w}(n+1) &= \overline{w}(n) + \mu X(n)[d^*(n) - x(n)\,\overline{w}(n)]\\ &= w(n) + \mu X(n)\varepsilon^*(n).\end{aligned} \tag{13}$$

Equation (13) is a continuous adaptive approach, where the weights are updated till the resulting weights converge to the optimum solution.

9.2.2   Handover algorithm in smart antenna systems

We noted above that in cellular systems, óne crucial consideration in call management is the transfer of a call from one BS to another when a mobile moves into the coverage area of a nearby base station. Conventional handover (or handoff) algorithms use signal strengths to decide if handover is necessary and which BS to hand over. However, the signal strength based algorithms become more complicated in smart antenna systems, since the signal strengths received will be a function of the radiation pattern of the BS antenna. Indeed a handover could be avoided by forming a new beam to the MS, so as to increase the uplink and downlink signal strengths. In addition, a non-control base station (NCBS) will receive weak signal strength from a nearby MS because that signal is considered as interference by the NCBS and hence a null will be put in the direction-of-arrival (DOA) of the MS. This also complicates the handover algorithm based on signal strength. Therefore we have proposed a simple distance-based handover algorithm for smart antennas. BS to MS distance is adopted as handover criteria since distance is the most direct parameter in deciding the control base station. It is understood that in areas where there is heavy shadowing, some form of signal strength criteria should be used with the nearest distance criteria.

Received signal strengths at three nearest BSs and the respective beam patterns formed at these BSs are used to find the position of the MS, with the CBS having the strongest received signal strength (see Section 9.2.3). The distance of the mobile from these three BSs is compared from time to time, and if the distance corresponding to the CBS is smaller than the distance between the MS and any of the other two BSs by a threshold of say, 10 m, handover will take place. Such a threshold is necessary to avoid extensive handover due to the fluctuations in the MS position at the edge of the cellular cell. Furthermore a triangle method is used to select the three nearest BSs from the MS by referring to the previous three nearest BSs.

9.2.2.1   Triangle method

The GSM system measures six signal strengths at any one time to decide the control base station. Once the transmission is being handed over to a new control base station, another set of six adjacent base stations will be chosen. Hence, from time to time, measurements from all the base stations within the same base station controller (BSC) coverage area will be compared. This procedure can be resource consuming since the number of BSs controlled by a BSC can be huge.

The estimated position derived from the position–velocity estimator is used to determine the three nearest BSs. Since any set of the three adjacent BSs forms a triangle, a new triangle will be formed when the MS (the user) moves out of the current triangle. However, only the third base station will be changed when a new triangle is formed. As shown in Fig. 9.3, BSs A and B will remain while BS C'will be replaced by BS C' when a new triangle is formed.

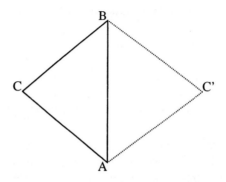

Figure 9.3: The old and the new triangle.

Hence, our objective is to find the new third adjacent BS when the MS (the user) moves out of the current triangle. In order to do that, we compute and compare three angles formed by the mobile user and the three adjacent base stations to check if the MS (the user) is out of the recently occupied triangle.

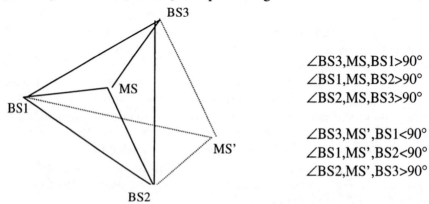

$\angle BS3,MS,BS1 > 90°$
$\angle BS1,MS,BS2 > 90°$
$\angle BS2,MS,BS3 > 90°$

$\angle BS3,MS',BS1 < 90°$
$\angle BS1,MS',BS2 < 90°$
$\angle BS2,MS',BS3 > 90°$

Figure 9.4: Angles involved in the triangle method.

As shown in Fig. 9.4, if the mobile user is inside the control triangle ($\angle$BS1, BS2, BS3), any two of the three angles shown must be larger than 90°. When any two of the angles are smaller than 90°, it is very likely that the MS is out of the control triangle. In addition, the further the MS is out of the triangle, the more likely that any two of the three angles will be smaller than 90°.

Once the MS is out of the "control" triangle, the following formula, with reference to Fig. 9.5, is used to calculate the rough position of the new third adjacent BS:

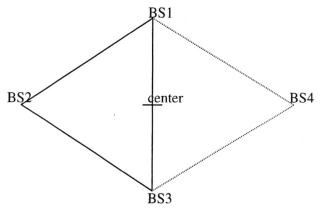

Figure 9.5: New third adjacent base station determination.

$$\text{center} = 0.5 \ (\ \text{BS1} + \text{BS3} \ ), \tag{14}$$

$$\therefore \quad \text{BS4} \ = \text{BS2} + 2 \ (\ \text{center} - \text{BS2})$$
$$= \text{BS2} + (\ \text{BS1} + \text{BS3} - 2 \ \text{BS2})$$
$$= \text{BS1} + \text{BS3} - \text{BS2}. \tag{15}$$

However, not all the BSs will have exactly the same separation (in our case study, approximately 1732 m). Hence, the calculated position of the new third adjacent BS is compared with all the existing BSs tied to a particular base station controller (BSC) or a mobile switching center (MSC), and the nearest BS will be chosen as the exact new third adjacent BS. Through this method, we only use the pre-determined positions of the existing BSs instead of getting time-varying information from all the BSs within the area of a particular BSC or MSC.

9.2.3   Position–velocity estimator (maximum likelihood method)

In smart antenna systems, an accurate MS positioning method is crucial for accurate beamforming. Besides, the information of the MS position is essential for the monitoring of the handover between base stations.

For our simulation purpose, we adopted the modified maximum likelihood method. From Section 3.9 we know that the radiated electric field strength from a finite length antenna element may be written in the following form:

$$E_n = E_0 \times \left[\frac{z_2 - z_j}{\sqrt{r_n^{\,2} + (z_2 - z_j)^2}} - \frac{z_1 - z_j}{\sqrt{r_n^{\,2} + (z_1 - z_j)^2}} \right] \times \text{AF}_n \qquad (16)$$

where E_n = electric field signal strength received at the BTS,

$$E_0 = 10^{-7}\omega\sqrt{\frac{P_r}{R_r}} \; ,$$

where ω = carrier frequency in rad/s, P_r = power radiated, R_r = radiation resistance of the transmitter antenna, z_2 = height of the BS antenna top, z_1 = height of the BS antenna bottom, z_j = height of the MS antenna centre, r_n = distance of MS from BTS, and AF_n = array factor.

Hence, the distance r can be estimated using

$$r_{n+1} = r_n + (Jr_n' Jr_n)^{-1} Jr_n' (E_m - E_n), \qquad (17)$$

where Jr = Jacobian matrix of the measured electric field strengths, E_m = column vector representing the measured electric field strengths at the BTS, and

$$Jr_n = \frac{\partial E_n}{\partial r_n}. \qquad (18)$$

The array factor for the smart antennas is given by

$$\text{AF}_n = \left| \text{conj}(W\_\text{opt}') \times \begin{bmatrix} 1 \\ e^{-j\Phi r} \\ e^{-j2\Phi r} \end{bmatrix} \right| \times N, \qquad (19)$$

$$\Phi r = \frac{2\pi d}{\lambda} \cos \Phi \sin \Theta,$$

$$\Theta = \pi - \tan^{-1}\frac{r}{z_1 + (L/4) - z_j},$$

$$\text{AF}_n = \left| \text{conj}(W\_\text{opt1}) + \text{conj}(W\_\text{opt2}) \times e^{-j\Phi r} + \text{conj}(W\_\text{opt3}) \times e^{-j2\Phi r} \right| \times N . \qquad (20)$$

The optimum weight W_opt is independent of the distance r between the MS and the BS. Thus for eqn (18) we have

$$Jr_n = J_1 + J_2, \qquad (21)$$

$$J_1 = -E_0 \left[\frac{z_2 - z_j}{\{r_n^{\,2} + (z_2 - z_j)^2\}^{3/2}} - \frac{z_1 - z_j}{\{r_n^{\,2} + (z_1 - z_j)^2\}^{3/2}} \right] r_n \times \text{AF}_n ,$$

$$J_2 = E_0 \left[\frac{z_2 - z_j}{\sqrt{r_n^2 + (z_2 - z_j)^2}} - \frac{z_1 - z_j}{\sqrt{r_n^2 + (z_1 - z_j)^2}} \right] \times \frac{\partial AF_n}{\partial r_n},$$

where

$$\frac{\partial AF_n}{\partial r_n} = \left| -j \frac{\partial \Phi r}{\partial r} \text{conj}(W\_opt2) e^{-j\Phi r} - j2 \frac{\partial \Phi r}{\partial r} \text{conj}(W\_opt3) e^{-j2\Phi r} \right| \qquad (22)$$

and

$$\frac{\partial \Theta}{\partial r} = -\frac{z_1 + L/4 - z_j}{r^2 + (z_1 + L/4 - z_j)^2}, \qquad (23)$$

$$\frac{\partial \Phi r}{\partial r} = \frac{2\pi d}{\lambda} \cos \Phi \cos \Theta \frac{\partial \Theta}{\partial r} \left(\cos \Theta = -\frac{z_1 + L/4 - z_j}{\sqrt{r_n^2 + (z_1 + L/4 - z_j)^2}} \right)$$

$$= \frac{2\pi d}{\lambda} \cos \Phi \frac{(z_1 + L/4 - z_j)^2}{\left[r_n^2 + (z_1 + L/4 - z_j)^2 \right]^{3/2}}, \qquad (24)$$

$$\therefore \frac{\partial AF_n}{\partial r} = \frac{2\pi d}{\lambda} \cos \Phi \frac{(z_1 + L/4 - z_j)^2}{\left[r_n^2 + (z_1 + L/4 - z_j)^2 \right]^{3/2}}$$

$$\cdot \left| j \, \text{conj}(W\_opt2) e^{-j\Phi r} + j2 \, \text{conj}(W\_opt3) e^{-j2\Phi r} \right|. \qquad (25)$$

After getting the value of r using eqn (17), the MS's position can be estimated using the formula given by eqn (27) in Chapter 7.

$$PM_{k+1} = PM_k + (Jp_k' Jp_k)^{-1} Jp_k' (0 - \text{errp}_k), \qquad (26)$$

where errp $= | \, PM - PB \, | - r$. Jp_k is defined as

$$\begin{bmatrix} \partial \text{errp}_1 / \partial x & \partial \text{errp}_1 / \partial y \\ \partial \text{errp}_2 / \partial x & \partial \text{errp}_2 / \partial y \\ \partial \text{errp}_3 / \partial x & \partial \text{errp}_3 / \partial y \end{bmatrix}.$$

From eqn (29) in Chapter 7, the velocity of the MS can be obtained from sequential PM estimates as follows:

$$V_m = (PM_m - PM_{m-1})/T, \qquad (27)$$

where T is the time interval between two discrete positions of the MS.

9.2.4   Channel model

9.2.4.1   Additive white Gaussian noise

In communication systems, noise is generated in many ways. We concentrate on the white noise, or the additive white Gaussian noise (AWGN). Further details of noise are found in Section 7.4.4. In this section we are concerned about the correlation of noise and the radiation (beam) pattern. The AWGN can be modeled by a normally distributed random value. For omni-directional transmission, AWGN is taken to be a zero mean value and has a variance of 0.161. In other words, with an SNR of 0 dB and signal power of 1 W, the noise power will be 0.161W, calculated from the formula below:

$$\text{Noise power} = \text{variance} \times \frac{\text{Signal power}}{10^{\text{SNR (dB)}/10}}. \tag{28}$$

However, the use of directive beams in smart antennas will affect the average noise power (or the noise variance) according to the beam pattern formed. This can be seen from Figs 9.6 and 9.7. Contrasting the beams shown in Figs 9.6 and 9.7, it can be seen that after beamforming, noises from certain directions are rejected or attenuated while those that fall within the beam pattern are amplified. The amplification of noise that falls within the narrow beam shown in Fig. 9.7 is due to the fact that within that area the array factor will have larger values (e.g. $AF = 3.0$) than that for omni-directional beams (e.g. $AF = 1.0$). Therefore, the noise power of the smart antenna system will no longer be the same as that of an omni-directional transmission system.

We propose that the noise power (or noise variance) received by the receiver is directly proportional to the surface area of the beam pattern:

$$\frac{\text{Noise variance}_{\text{BF}}}{\text{Beam surface area}} = \frac{\text{Noise variance}_{\text{isotropic}}}{\text{Sphere surface area}} = \frac{0.161}{4\pi}, \tag{29}$$

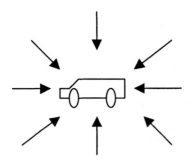

Figure 9.6: Noise from all directions.

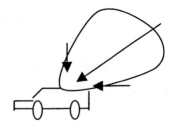

Figure 9.7: Noise after beamforming.

$$\text{Noise variance}_{BF} = \text{Noise variance}_{isotropic} \times \frac{\text{Beam surface area}}{4\pi}. \tag{30}$$

We define noise factor as the ratio of noise power after beamforming (BF) to that before beamforming (i.e. omni-directional):

$$\text{Noise factor} = \frac{\text{Noise variance}_{BF}}{\text{Noise variance}_{isotropic}} = \frac{\text{Beam surface area}}{4\pi}. \tag{31}$$

We shall illustrate now how the noise factor may be calculated. Consider a three-element ($N = 3$) array antenna. We want to determine its noise factor by first evaluating its beam surface area using a formula. The array factor is given by

$$AF = \left| (w_{1r} + jw_{2i}) + (w_{2r} + jw_{2i})e^{j\Phi r} + (w_{3r} + jw_{3i})e^{j2\Phi r} \right|, \tag{32}$$

where w_r = real part of the respective weight component and w_i = imaginary part of the respective weight component,

$$\Phi r = \cos\left(\frac{2\pi d}{\lambda} \sin\Theta \cos\Phi \right). \tag{33}$$

Hence

$$AF = \sqrt{A\cos 2\Phi r + B\cos \Phi r + C + D\sin 2\Phi r + E\sin \Phi r}, \tag{34}$$

where

$$A = 2(w_{1r}w_{3r} + w_{1i}w_{3i}),$$
$$B = 2(w_{1r}w_{2r} + w_{1i}w_{2i} + w_{2r}w_{3r} + w_{2i}w_{3i}),$$
$$C = w_{1r}^2 + w_{1i}^2 + w_{2r}^2 + w_{2i}^2 + w_{3r}^2 + w_{3i}^2,$$
$$D = 2(w_{1i}w_{3r} - w_{1r}w_{3i}),$$
$$E = 2(w_{1i}w_{2r} - w_{1r}w_{2i} + w_{2i}w_{3r} - w_{2r}w_{3i}).$$

Therefore, the surface area of the radiation pattern is given by

$$\text{Beam surface area} = \int_{\Phi=0}^{2\pi} \int_{\Theta=0}^{\pi} |AF|^2 \sin\Theta\, d\Theta\, d\Phi. \tag{35}$$

An approximate method to get the noise factor is described in the following steps:

Step 1 – Generate a number of noise channels (in practice, noise channel number is infinite), say 1000, with random direction-of-arrival(DOA):

$$x(j) = \sqrt{\text{variance}} \times \text{random\_number\_normal}, \tag{36}$$
$$\text{rand\_}\Theta(j) = \pi/2 + \text{random\_number} \times \pi, \tag{37}$$
$$\text{rand\_}\Phi(j) = \text{random\_number} \times \pi, \tag{38}$$

where variance = 0.161, random_number_normal is normally distributed with zero mean value and has a variance of 1, random_number is uniformly distributed within 0 and 1, rand_Θ = angle between the noise channel's direction and the z-axis, rand_Φ = azimuth angle of the noise channel's direction, and j = 1, 2, ..., 1000.

Step 2 – Evaluate the average noise power (without beamforming), N_isotropic:

$$N\_\text{isotropic} = \frac{1}{1000} \sum_{j=1}^{1000} x(j)^2. \tag{39}$$

It should be approximately 0.161 (variance of the noise channel). This serves as a means to verify whether the noise channel number used is sufficient to simulate the infinite number of noise channels.

Step 3 – Use the same noise channels and multiply the noise signal, $x(j)$, with their respective array factor, AF(I), according to the DOA of each noise signal.

Step 4 – Evaluate the average noise power received after beamforming, N_BF:

$$N\_\text{BF} = \frac{1}{1000} \sum_{j=1}^{1000} [\text{AF}(j)x(j)]^2. \tag{40}$$

Step 5 – Calculate the ratio of the noise power level after beamforming to the noise power level in the isotropic scenario that is the noise factor.

$$\text{Noise factor} = \frac{N\_\text{BF}}{N\_\text{isotropic}}. \tag{41}$$

9.2.5   Performance evaluation

In this section we discuss the analytical tools and computer simulations to evaluate the performance of a CDMA cellular system using smart antennas. The results and simulations used in the performance evaluation are discussed in the later part of this chapter.

9.2.5.1   System capacity
We may model a hexagonal cellular system by two concentric circles, with the outer circle being subdivided into six segments (neighboring BSs). The two concentric circle model and the six outer BS segments are shown in Fig. 9.8. This is due to the difficulty in generating the scenario for the performance evaluation module of the CDMA system when using a hexagonal system.

Interference of the user in adjacent cells. When an MS is transmitting a signal to its control BS (the CBS), it is also transmitting a signal to all the neighboring BSs as well. The signal will be treated as an interferer by the neighboring BSs. The magnitude of the signal received by the neighboring BSs will be smaller than that received by the CBS. This may be due to path loss and the geometry of the MS beam which may seek to maximize transmission to its own CBS. The identity of the CBS will change as handover occurs with the movement of the MS.

Let us assume that the signal power of the MS received by the CBS is P_0. In order to find the signal power received by a neighboring cell, the distance between the MS and

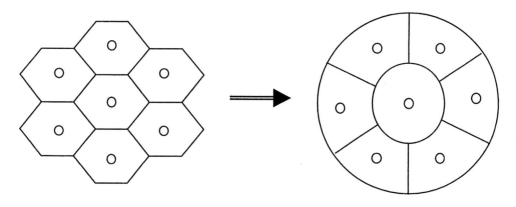

Figure 9.8: Modeling of the cellular system.

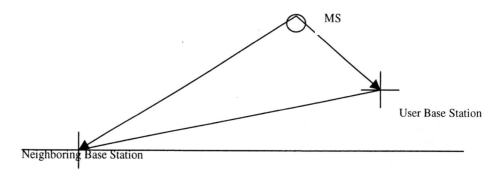

Figure 9.9: MS and adjacent BSs.

the neighboring BS must be known. The distance can be easily found using cosine rule (see Fig. 9.9).

Let the angle-of-arrival of the MS be Θ_a with distance R_0 from the user base station (CBS). Further, the angle subtended by the CBS to the neighboring base station is Θ_b with the distance between them being $2R$.

Let $P_{user,bs}$ and $P_{user,\,ns}$ be the power received from the CBS and the neighboring BS respectively and P_{tx} be the power transmitted by the MS. Using the cosine rule, the distance between the MS and the neighboring BS is obtained from

$$D^2 = R_0^2 + (2R)^2 - 2(R_0)(2R)\cos(\pi - \Theta_a + \Theta_b). \qquad (42)$$

We shall assume (a) perfect power control, (b) the CBS is required to receive a minimum power of P_{req} and (c) the *log distance path loss model*. Hence the power received at the CBS is given by

$$P_{user,bs} = G_{tx} \times G_{rx} \times P_{tx} \times \left(\frac{\lambda}{4\pi d_{ref}^2}\right)^2 \left(\frac{d_{ref}}{R_0}\right)^n \tag{43}$$

$$= G_{rx} \times P_{tx} \times \left(\frac{\lambda}{4\pi d_{ref}^2}\right)^2 \left(\frac{d_{ref}}{R_0}\right)^n, \tag{44}$$

where G_{tx} = isotropic gain of transmitter antenna = 1, G_{rx} = BS receiver gain, d_{ref} = reference distance for path loss calculation and n = path loss exponent. We may rewrite eqn (44) as

$$P_{user,ns} = G_{rx} \times P_{tx} \times \left(\frac{\lambda}{4\pi d_{ref}^2}\right)^2 \left(\frac{d_{ref}}{D}\right)^n. \tag{45}$$

Therefore, the out-of-cell interference by a single MS user is given by

$$P_{outcell\text{-}user} = P_{pc} \times \left(\frac{R_0}{D_n}\right)^n$$

$$= P_{user,bs} \times \left(\frac{R_0}{D}\right)^n, \tag{46}$$

where n is the path loss exponent factor and it is dependent on the specific propagation enviroment, and P_{pc} is the minimum power required by the CBS in order to achieve perfect power control. Table 9.1 lists the typical path loss exponents for various mobile radio environments.

Since every cell BS smart antenna is being beamformed, the signal to the CBS might be attenuated relative to the maximum array factor amplitude of the cell antenna. Thus in order to achieve perfect power control, each MS has to transmit more power in order for its CBS to receive the required signal strength. Therefore eqn (46) has to be modified as follows:

$$P_{outcell\text{-}user} = P_{pc} \times \left(\frac{R_0}{D}\right)^n \times \left(\frac{AF_{max}}{AF_{bs,user}}\right)^2$$

$$= P_{pc} \times \frac{(R_0/D)^n}{AF_{bs,user,norm}^2}. \tag{47}$$

Due to the CBS antenna beamforming, the power will be further reduced by multiplication of the array factor and the interference signal power. Therefore the out-of-cell interference of a single MS is

$$P_{\text{outcell-user}} = P_{\text{pc}} \times \frac{(R_0/D)^n}{\text{AF}^2_{\text{bs,user,norm}}} \times \text{AF}^2_{\text{ns,user}}. \tag{48}$$

In our simulation studies, we generated K numbers of MSs with random DOA and distance in each of the six adjacent cells with random distance and angle from its base station. Thus, the total outer cell interference power can be calculated from the summing of the transmission power from all the MSs in the six cells.

Table 9.1: Path loss exponents in wireless communications environment.

Environment	Path loss exponent
Free space	2
Urban area cellular radio	2.7–3.5
Shadowed urban cellular radio	3–5
In-building line-of-sight	1.6–1.8
Obstructed in-building	4–6
Obstructed in factories	2–3

Therefore $6 \times K$ users from all the adjacent cells will contribute to the out-of-cell interference. In other words, the total out-of-cell interference will be

$$P_{\text{outcell}} = 6 \times K \times P_{\text{outcell-user}}$$

$$= 6 \times K \times P_{\text{pc}} \times \frac{(R_0/D)^n}{\text{AF}^2_{\text{bs,user,norm}}} \times \text{AF}^2_{\text{ns,user}}$$

$$= 6 \times K \times P_{\text{pc}} \times \beta,$$

where

$$\beta = \frac{(R_0/D)^n}{\text{AF}^2_{\text{bs,user,norm}}} \times \text{AF}^2_{\text{ns,user}}.$$

Interference from MSs in the cell under study. Suppose there are K MS users in the cell under study. Then the single MS under study will treat the other $K-1$ MSs as interferers. Since perfect power control is used in the CDMA system, the power of all the MSs received in the cell under study will be the desired power, P_{pc}. Due to BS smart antenna beamforming in the cell under study, those MSs which are not in the direction of the main lobe of the radiation pattern will need to transmit additional power in order to achieve the power of P_{pc}.

Therefore, the total in-cell interference power will be

$$P_{\text{incell}} = (K-1) \times P_{\text{pc}}. \tag{49}$$

From the generation of the interferences for both the in-cell and out-of-cell MSs, we can calculate the reuse factor as follows (Liberti and Rappaport, 1999):

$$\text{Reuse factor} = \frac{P_{incell}}{P_{incell} + P_{outcell}}$$

$$\approx \frac{1}{1+6\beta}$$

$$= \frac{(K-1)P_{pc}}{(K-1)P_{pc} + 6KP_{pc}\beta} \qquad \text{(for } K>>1\text{).} \qquad (50)$$

Carrier to Interference and Noise Ratio (CINR) is defined after spreading as the ratio of the desired signal to the sum of interference and noise, i.e.

$$\text{CINR} = \frac{P_0}{(1/N)\sum_{k=1}^{K-1} P_k + \sigma_n^2}, \qquad (51)$$

where P_0 is the power of the desired signal at the input of the despreader, P_k is the power from the other user, σ_n^2 is the noise contribution after despreading and

$$N = \text{Processing gain} = \frac{\text{Chip rate (chip per second)}}{\text{Information symbol rate(symbol per second)}}. \qquad (52)$$

After despreading, the noise bandwidth is $1/T_b$. The CINR can be rewritten as

$$\text{CINR} = \frac{P_0 T_b}{(1/N)\sum_{k=1}^{K-1} P_k T_b + \sigma_n^2 T_b}$$

$$= \frac{E_b}{N_i + N_n}$$

$$= \frac{E_b}{N_0},$$

where N_n is the power spectral of thermal noise and N_i is the power spectral density of the total MAI after despreading.

In our simulation, we can express CINR as

$$\text{CINR} = \frac{NG_a P_c}{v(K-1)P_c + 8vKP_c\beta}$$

$$= \frac{NG_a}{v(K-1) + 8vK\beta}$$

$$\approx \frac{NG_a}{vK(1+8\beta)}. \qquad (53)$$

Rearranging, the capacity of the CBS can then be evaluated as shown:

$$\text{Capacity} = \frac{\text{Reuse factor} \times \text{Processing gain } (N) \times \text{Pattern gain } (G_a)}{\text{Voice activity } (v) \times E_b/N_0}, \tag{54}$$

where

$$\text{Processing gain} = \frac{\text{Chip rate (chip per second)}}{\text{Information symbol rate (symbol per second)}}.$$

The voice activity (v) factor arises due to the fact that the MS will not be transmitting continuously, when the speaker pauses in a conversation over the mobile phone (MS). The vocoder will reduce the output rate when the speaker is silent. Pattern gain is the smart antenna horizontal azimuth gain after beamforming. The ratio E_b/N_0 is the one required by the receiving BS, i.e. the CBS.

9.2.5.2 Loading of antenna

The fully adaptive array antenna (i.e. the smart antenna) is capable of maximizing its beam array factor in the direction of the signal of interest while nulling the beam in the directions of the interferers. However, there is a limitation to the number of interferers that can be nulled due to the antenna array size (i.e. the number of antenna elements). We shall now evaluate the dependence of the number of interferers that the antenna can null on the array antenna size.

9.2.5.3 Signal to interference and noise ratio (SINR)

The smart antennas can effectively cancel out the interference in the system if the system is not overloaded. Therefore in a practical case, not all the interferences can be nulled and thus some interferers will contribute to the SINR. In our simulation studies we shall evaluate SINR for the whole system and find out how much SINR improvement can be achieved by using smart antennas comparing its performance with a conventional antenna.

SINR is defined as

$$\text{SINR} = \frac{\text{Signal power}}{\text{Total interference} + \text{noise}}. \tag{55}$$

9.2.5.4 Range

A BS smart antenna is expected to provide an increase of range that the BS can cover. The improvement in smart antenna range can be assessed by comparing it with the distance coverage by a conventional antenna system. The antenna gain, which can be numerically determined using Trapezium rule from the surface area of the beam pattern, is used to evaluate the range improvement. One of the important factors that affect the range of a system is the loss exponent. In other words, the system will exhibit maximum range in free space and the range will be degraded in a built-up urban area.

9.3   Simulation studies

9.3.1   Simulation scenario

The overall simulation is based on the scenario used in Chapter 7. There are 16 base stations (BS, marked as "A" in Fig. 9.10) in the scenario with a cell radius of 1000 m. A mobile station (MS, marked as "x" in Fig. 9.10) travels along the dotted path, with the velocity variation shown in Fig. 9.11.

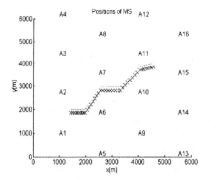

Figure 9.10: BS and MS positions.   Figure 9.11: MS velocities versus
                                                                    sampling time.

Forty sampling points are used with a time interval of 4.8 s (equivalent to 10,000 sampling intervals adopted in the GSM system time step of 0.48 ms). Some of the para+meters used are shown in Table 9.2.

Table 9.2: Simulation parameters.

Parameters	Description
$f = 1250$ MHz	Carrier frequency
$z_1 = 60$ m	Height of the lower part of the BS antenna
$z_2 = z_1 + \lambda/2$	Height of the upper part of the BS antenna
	($L = \lambda/2$ for half-dipole antenna)
$z_j = 1$ m	Height of the MS antenna
$R = 73\ \Omega$	Radiation resistance of the MS antenna
$P = 2$ W	Radiation power of the MS antenna

9.3.2   Algorithm

The simulation proceeds as follows. The computer program is contained on disk (Program 9.1).

1. Initialization;
2. Choose CBS and BS (1:3),

(a)   get estimation of distances r (initial value for maximum likelihood estimator's iteration) using time advanced (TA) measurement.

(b)   choose CBS by referring to the *estimate_r*. BS (1:3) refers to the three BS with the smallest *estimate_r*;

(c)   generate 1st set of θ and Φ using exact parameters $(r, z1, \dots)$;

(d)   generate parameters (*rand_θ, rand_interf, rand_Φ*) for random beam patterns for all BSs (*rand_θ* calculation based on θ); *rand_interf* is the random azimuth angle of the interference (assuming only one interferer);

(e)   generate array factors of all the BS with respect to the signal of interest (SOI) - MS, using parameters generated in 2(c) and 2(d). (call function *beamform.m*);

(f)   generate E-fields received by all the BS (without noise);

(g)   calculate E_m (with noise);

(h)   choose CBS: In case there are more than one BS (within the three nearest BS) with the smallest *estimate_r* (noting that there is an estimation error of TA of 277 m), CBS is chosen to be the BS with the strongest received E_m;

3. Sampling loop (40 points),

   (a) $kk = 1$;

   (b)   if $(kk>1)$,

       (I)   generate θ and Φ using exact parameters $(r\_used, zj\_used, \dots)$;

       (II)   generate array factors of the three nearest BS with respect to the SOI (call function *beamform.m*);

       (III)   calculate E_m_used (which is the E_m of the three nearest BS);

   (c)   iteration of *ddr* (iterated value of r);

   (d)   iteration of *pp* (iterated position of MS);

   (e)   if $(kk>1)$, get velocity;

   (f)   prepare parameters for the next sampling point;

   (g) $kk = kk + 1$;

   (h)   go back to (b).

3c.   Iteration of *ddr*,

   (I) $k = 1$;

   (II)   set initial value of *ddr* as the *estimate_r*;

   (III)   generate $d\theta$ using *ddr*, assuming Φ is known to the system $(d\Phi = \Phi)$;

   (IV)   generate respective AF_used (1:3) and W_opt (1:3) using $d\Phi$ and $d\theta$ (call function *beamform.m*);

   (V)   generate EE (*estimate_E*) and J (Jacobian matrix of EE) (call function *jascob.m*);

   (VI) $ddr = |ddr + (J^T J)^{-1} J^T (Em\_used - EE)|$;

   (VII) $k = k + 1$;

   (VIII)   stop r estimation if rcond(J) ≤ 0.01 (reciprocal condition estimator) (prevent Jacobian matrix from becoming singular);

   (IX)   go back to (III).

3d.   Iteration of *pp*,

   (I) $k = 1$;

(II)   set initial value of pp (MS position) as the center of the three nearest BS;

(III)   shift pp by 10 m in x-y direction when pp is too near (0.1 m) from any of the three adjacent BS;

(IV)   evaluate $Errp$ ($= |PB - pp| - ddr$), the position error;

(V)   evaluate Jp, the Jacobian matrix of $Errp$;

(VI)   stop position estimate when cond(Jp) > 500;

(VII) ppx (pp in x-direction) $= ppx + (Jpx^T \times Jpx)^{-1} \times Jpx^T \times (-Errpx)^T$;

$\qquad$ ppy (pp in y-direction) $= ppy + (Jpy^T \times Jpy)^{-1} \times Jpy^T \times (-Errpy)^T$;

(VIII) $k = k + 1$;

(IX)   go back to (III);

3e.   Get velocity of the MS,

$\qquad$ $V_t = (PM_t - PM_{t-1}) / T$;

3f.   Prepare parameters for the next sampling point,

(I)   deciding on handover by using ddr :

$\qquad$ handover occurs when ddr(old_CBS) - ddr(new_CBS) > 10 m. A tolerance of 10 m is created to avoid excessive handover while the PM fluctuates between two cells;

(II)   decide whether or not to change BS smart antenna beam by simulating a beamformed scenario and compare the E-field received by the CBS before and after beamforming : a new beam that is steered towards the MS will be formed when a significant improvement in received E-field is expected;

(III)   generate random beam patterns for the two adjacent BSs (exclude CBS), by specifying $rand\_\Phi$ and $rand\_interf$;

(IV)   decide on a new set of adjacent BSs using triangle method;

(V)   update all other parameters (e.g. PM, PB_used, ...) for the next sampling point;

(VI)   set the iterated r value (ddr) as the initial value for the next r-iteration;

9.4   Results and discussion

9.4.1   BS smart antenna beams

Figure 9.12 shows the effect of the number of antenna elements on the smart antenna radiation pattern or beam formed. As expected, the beam becomes narrower, more directive and better focused on the MS as the number of antenna elements is increased.

The results shown in Fig. 9.12 are for an interfering signal which is at a large angular distance from the desired signal. The result shows that the interfering signal will be nulled regardless of the number of elements of the smart antenna. Canceling interference is one of the objectives in using a smart antenna. As the number of elements in the array antenna is increased, the radiation pattern will have more sidelobes. The sidelobe levels are much smaller when compared to the pattern of a smaller array. Hence an array with more elements will minimize the undesired signals, including multipaths. Hence the effect of a large array is to increase the output signal to noise and interference ratio. In addition, the direction in which the peak point of the radiation pattern points will more closely correspond to the direction of the desired signal.

N= number of antenna elements
Angle of arrival of desired signal : $45°$
Angle of arrival of undesired signal : $135°$

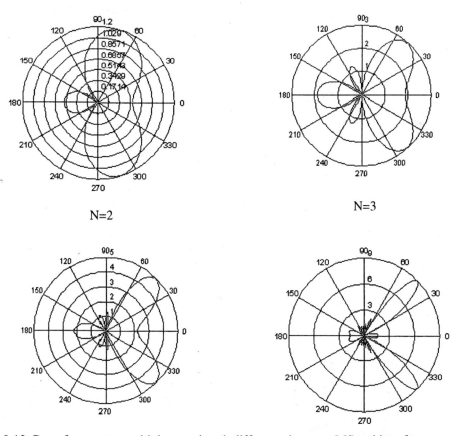

Figure 9.12: Beamform pattern with large azimuth difference between MS and interferer.

The array factor (AF) of the smart antenna in the direction of the desired signal is very much higher than for an isotropic antenna. This is seen from Table 9.3. For a nine-element smart antenna, the AF in the direction of the desired signal is 8.9495, much higher than the AF= 1.0 we would have obtained with an isotropic antenna. In addition to this advantage, the AF is cut down to 0.0064 (from 1 for isotropic antenna) in the direction of the undesired signal. The half-power beam width (HPBW) is 20° and the side-lobe-level (SLL) is 1.8889. The beam width for a bigger array antenna will become narrower. If the beam width is too small, the MS may very rapidly fall out of the main lobe if the MS is traveling at a high speed. Thus there is a need to arrive at a compromise between the beam width and the accuracy of the beam or radiation pattern.

The smart antenna was also tested for an interfering signal that is angularly near to the desired signal. The beam obtained was not as accurate as for the previous test with the same number of elements. This is due to the fact that the antenna is unable to null the interference and point to the desired signal with the restricted degree of freedom, which is equal to the number of elements. This observed degradation of performance will be important when implementing cell sectorization. Cell sectorization will reduce the number of undesired signal for each smart antenna in any one sector. However, if there is any undesired signal, it means that the undesired signal will be angularly close to the desired signal.

Table 9.3: Radiation pattern characteristics with smart antenna.

N, number of elements	2	3	5	9
$AF_{desired}$, array factor in the direction of MS	1.9250	2.9496	4.9473	8.9495
$AF_{undesired}$, array factor seen by interferer	0.0443	0.0077	0.0130	0.0064
HPBW, half-power beam width	80°	52°	38°	20°
SLL, side-lobe-level	1.0989	1.8861	2.2085	1.8889

9.4.2   Triangle method

Table 9.4 shows the location of the three BSs nearest to a moving MS. The SNR was 10 dB. In Table 9.4 *kk* is the sampling number, ranging from 1 to 40; adjacent BS number indicates three nearest BS numbers ranging from 1 to 16 (refer to Fig. 9.10); *ddr* is the computed distance of MS from adjacent BS.

Tests were also performed for SNR = 20 dB. The tests performed verified the usefulness of the triangle method. The change of the third nearest BS (shown in bold in Table 9.4) occurs at almost the same sampling number (*kk*) for all the tests performed at different SNR. It can be seen that a new third nearest base station will be "computed" only when more than one angle is less than 90° (the underlined angle). However, some complex angles (shown in italics) are generated due to the estimation error in *ddr*. This erroneous outcome is because of taking the arc cosine of a value bigger than "1". Even though the occurrence of these complex angles is rare (approximately 5%), there exists a need to further modify the triangle method.

9.4.3   Handover

Figure 9.13 and Table 9.5 show how handover takes place as the MS moves through a region installed with 16 BSs with smart antennas. In Fig. 9.13, B is the control base station (CBS), and O is the position of the MS where handover occurred. The dotted

Table 9.4: Triangle method for locating the three nearest BSs.

Kk	Adjacent BS no.	ddr	New adjacent BS no.	Computed PM (m)	Angle(kk) (°)	Angle(kk−1) (°)
12	1, 2, 6	1954, 1227, 518	**7**, 2, 6	(2261, 2407)	<u>61</u>, 165, <u>57</u>	80, *180−j27*, 105
18	7, 2, 6	635, 1734, 999	7, **10**, 6	(2692, 2920)	<u>79</u>, <u>73</u>, *180−j40*	96, <u>80</u>, *180−j25*
27	7, 10, 6	1228, 634, 1668	7, 10, **11**	(3592, 3153)	134, <u>85</u>, <u>71</u>	149, 94, <u>80</u>
34	7, 10, 11	1686, 1294, 616	**15**, 10, 11	(4183, 3944)	<u>70</u>, 126, <u>84</u>	<u>75</u>, 144, 98

SNR=10 dB, noise type=AWGN
Seven complex angles out of 117 computed angles.

track in Fig. 9.13 corresponds to the estimated positions of the MS and the dashed track indicates the exact path of the MS.

Further tests at different SNR indicated that all the handovers take place almost at the same sampling number irrespective of the SNR of the entire region. No experiments were carried out for SNR variations within the area covered by the 16 BSs. The handover algorithm can be improved by taking into account the electric field or power strength received by the three nearest BSs.

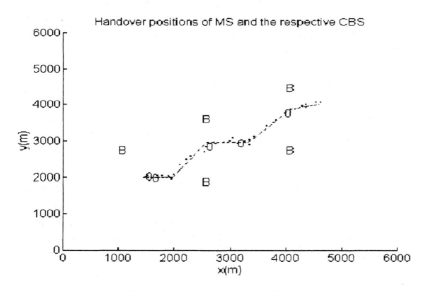

Figure 9.13: Handover of MS control

Table 9.5: Handover of MS control. The CBS is highlighted.

Sampling no.	*ddr*	*ddr*(Sampling no. – 1)
3	1251, 983, **940**	1109, **893**, 995
16	**772**, 1475, 1046	864, 1376, **773**
23	1007, **962**, 1141	**802**, 986, 1235
31	1472, 900, **875**	1356, **806**, 999

SNR = 10 dB

9.4.4. BS-based position–velocity estimator

In Chapter 7 we discussed MS-based position–velocity estimation of the MS using electric field strengths of signals from the BS picked up at the MS. Figure. 9.14 shows the BS-based position–velocity estimation results for SNR = 20 dB using electric field strengths of signals from the MS picked up at the BS. With an SNR of 20 dB, the average position estimation error is approximately 23 m. Besides, it is also shown that all estimated positions are accurate to 70 m. The velocity estimator achieved an average velocity estimation error of 24 km/h.

For SNR = 10 dB, it was found that the position estimator achieved quite satisfactory, with an average estimation error of 78 m, which is much better compared to the estimation error of 277 m with the time delay method. Besides, it is also shown that all the position estimation errors using BS base position estimation using electric field measurements are below 200 m. In other words, all positions are successfully estimated.

However, the velocity estimator for SNR = 10 dB was quite poor. The velocity estimation error of almost 100% was observed at some point, and this is simply intolerable. This error is accumulated from the position estimation error on which the velocity estimator depends.

9.4.5. AWGN model for smart antenna systems

Table 9.6 shows the effect of interference on the array factor of the smart antenna. From the data shown in Table 9.6, it is seen that the noise factor of an AWGN channel after beamforming is bigger than 1. In other words, the channel noise power (or variance) is increased after a beam is formed – an unexpected result. This can be explained by observing the beam shown in Fig. 9.15. The beam, or radiation pattern, shown in Fig. 9.15 is for the case when the desired signal is at 131.5° and interference comes from 108.0°. Observe that the beam pattern formed is bigger than for the isotropic case where we get a sphere with unity array factor. Therefore the noise factor tends to increase with beamforming due to the increase in beam surface area that could receive more noise power.

SNR = 20 dB

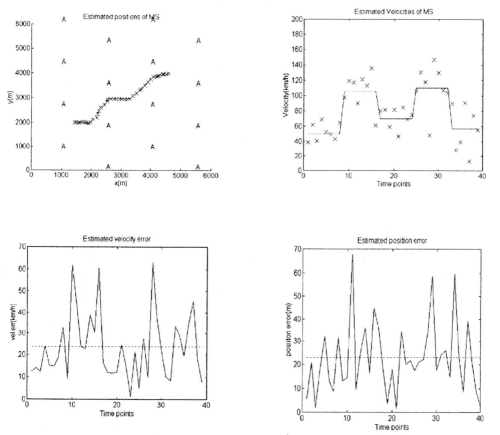

Figure 9.14: Position–velocity estimation with BS-smart antennas.

Table 9.6: Effect of radiation pattern area on noise at the receiver.

Signal (θ, ϕ)	Interference (θ, ϕ)	Noise factor	Area/(4π) of beam	Array factor^2
91.7, 43.4	91.7, 143.8	2.85, 2.99, 2.69, 2.89, 3.22	3.42	8.65
91.7, 24.9	91.7, 139.4	4.27, 4.27, 4.03, 4.32, 4.31	4.93	8.49
91.7, 131.5	91.7, 108.0	4.17, 4.13, 3.79, 4.04, 4.20	4.62	8.53
91.7, 154.0	91.7, 11.4	16.31, 15.87, 16.38, 15.86, 15.46	17.90	6.77
91.7, 54.1	91.7, 86.8	2.78, 2.70, 2.48, 2.50, 2.41	3.11	8.69
91.7, 170.0	91.7, 80.9	2.27, 2.21, 2.08, 2.13, 2.24	3.11	6.61

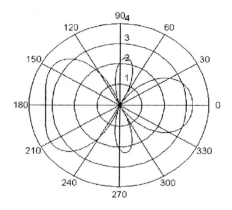

Figure 9.15: Illustrating increase in noise factor.

However, the increase in noise factor does not mean that the beamforming will impair the signal transmission. As the signal power is proportional to the square of the array factor, it can be seen that the increase in signal power is much bigger than the noise factor. Therefore the overall effect of beamforming will be to increase the signal to noise ratio.

Besides, it is verified that the noise factor is also equal to the area ratio beam area/(4π). In other words, the bigger (in surface area) the beam that is formed, the larger the value of the noise factor. The small discrepancy between the noise factor and the area ratio is due to the approximation error introduced while computing the beam area numerically, and the simulation error created while generating noise factor.

For the case when the signal is at ($91.7°$, $154°$) and interference comes from ($91.7°$, $11.4°$), it is seen from Table 9.6 that the noise factor is larger than the (array factor)^2 of the signal. This is due to the fact that we did not consider the smart antenna being used in a sectored form. Even though the beam is not exactly symmetrical at the $90–270°$ axis (it is exactly symmetrical along the $0–180°$ axis), the algorithm will not be able to create a large array factor at ϕ when a null is expected at ($180°-\phi$). In these unfavorable circumstances, a large beam will be formed resulting in a large noise factor. But the value of the array factor for the desired signal is still relatively small. This situation can be avoided if sectioning is used. Figure 9.16 shows a sectored arrangement of smart antennas. Figure 9.16(a) shows a BS arrangement with three sectors. Each sector has a smart antenna with five elements that will cover $120°$ each. Any sidelobe or backlobe radiation in the backward direction will be absorbed to prevent interference with the antennas of the other sectors. Figure 9.16(b) shows a three-sectored BS arrangement, where each sector is divided into two separate arrays so as to allow for better diversity and hence better noise or interference cancelation.

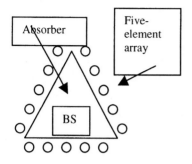

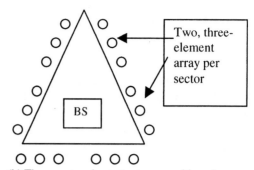

(a) Three-sectored smart antenna

(b) Three-sectored smart antennas with each sector having two, three-element arrays for diversity.

Figure 9.16: Sectored smart antenna arrangements.

9.4.6   Performance evaluation

9.4.6.1   Capacity, SIR and range

The simulation is tested for the route shown in Fig. 9.10 having 40 discrete positions. Each BS is assumed to have three-element arrays. Figure 9.17 shows the range, capacity and SIR performance of the three-element BS antennas compared to isotropic BS antennas. Sectoring, which will improve the performance of the smart antenna further, was not used. With BS beamforming performed by a simple three-element smart antenna, it can be seen from Fig. 9.17(a) that the BS can cover a longer range.

It can also be seen that with beamforming of the base stations, the range (Fig. 9.17(b)) and SIR (Fig. 9.17(c)) are improved when compared to an isotropic antenna. These observations make the use of BS smart antennas attractive. The simulation results reveal that for 10 users in each cell in an urban environment ($n = 4$), a mean range improvement of 1.2799 dB, and a mean capacity of 41 MS users and an SIR improvement of 2.9394 could be obtained. However, the simulation results shown are based on the assumption that perfect power control is achievable.

The limitation of BS beamforming in multiple directions may demand an increase in the MS transmitted power if an MS is not positioned at the peak of the BS antenna beam. This may result in a shortened battery life. To overcome this the BS may be sectored into three sectors, each covering $120°$ as shown in Fig. 9.16(a). Depending on the position distribution of the users in the cell, the BS smart antenna must maximize its beam pattern for all the MS users in the cell in order to minimize the transmission power required of the MS user.

It was shown in Section 9.4.1 that HPBW will decrease as the number of antenna elements increase . Thus, the BS antennas can be designed as follows:

1. Each cell is sectored to $120°$ each.
2. Each sector will have a reasonably large array so that a narrower beam can be formed.

3. Each sector is again sectored into eight subsectors (for simplicity), each covering 15°.

The advantage of sectoring is that the array factor will be large for all MS users instead of a few that might be clustered around a single BS antenna beam peak. This means the required MS user transmission power will be less. However, the disadvantage is that such sectoring is only possible where MS users are usually clustered within a few subsectors not evenly distributed in the sectors. Handover will become more complicated with sectoring.

9.4.6.2 Loading of antenna

In this section we study the effect of interferers on the loading of the smart antenna and its accuracy in generating the desired radiation pattern.

Figure 9.18 shows the changes in the BS smart antenna beam when the number of interferers increases. It is shown that as the number of interferers increases, some of the interferers are not being nulled by the adaptive smart antenna system. This is because an array antenna system of N number of elements can only effectively null $N-1$ number of interferers. In addition, the direction of the main beam will be slightly shifted if the direction of any interferers is too close to the direction of the desired signal. This will tend to cause tthe gain in the direction of the desired signal to degrade.

Table 9.7 shows the interference power level with reference to the desired signal power level at the receiver electronics using a four-element array antenna. The desired signal is at 60°. It is obvious that the performance of the smart antenna will depend on the number and position of the interferers. As the number of interferers increases the interference power level approaches that of the desired signal power.

Table 9.7: Relative interference power level at the receiver as interferers increase in number using a four-element antenna.

Angle of interference (°)	Total number of interferers			
	2 (dB)	3 (dB)	4 (dB)	5 (dB)
30	−46.03	−42.74	−20.11	−18.86
90	−60.55	−50.39	−32.60	−22.94
135		−44.08	−21.04	−14.96
150			−16.43	−13.25
45				−8.55

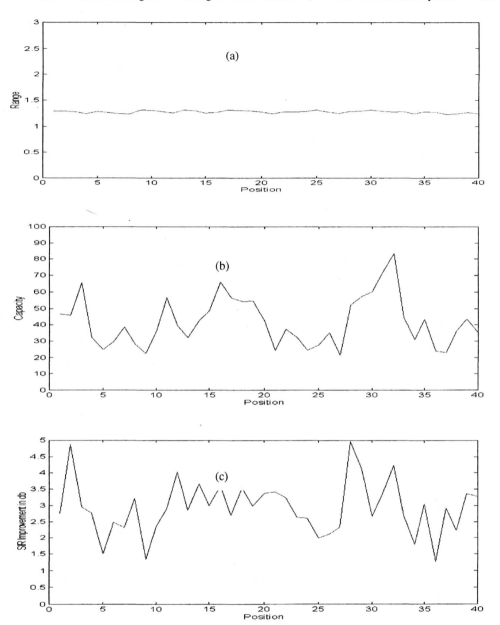

Figure 9.17: Performance improvement with smart antennas.

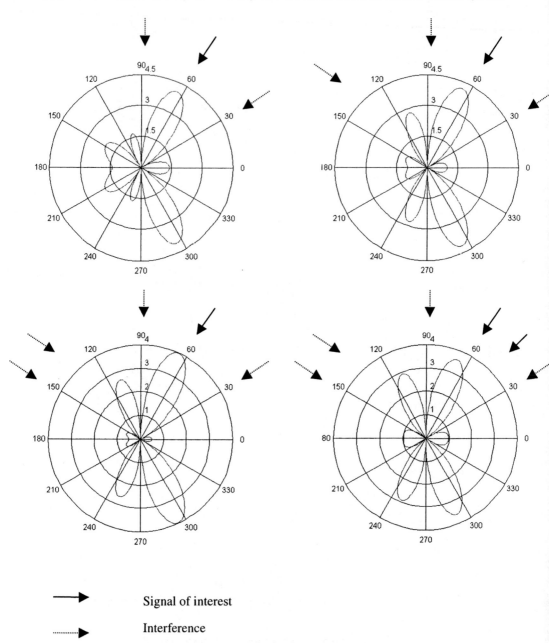

Figure 9.18: Antenna loading due to interferers

Bibliography

In writing this book, the books and papers listed below were referred to. The reader should consult them for further details or for a different perspective on the subject.

M.D. Austin, Velocity adaptive handoff algorithms for microcellular systems, *IEEE Trans. Vehicular Technology*, **43**, 549–560, 1994.

C.A. Balanis, *Antenna Theory*, 2nd edn, John Wiley, New York, 1997.

W.G. Carrara, R.S. Goodman and A. Majewski, *Spotlight Synthetic Aperture Radar Signal Processing Algorithms*, Artech House, Boston, 1995.

L. Castedo, An adaptive beamforming technique based on cyclostationary signal properties, *IEEE Trans. Signal Processing*, **43**, 1637–1650, 1995.

Chan Teck Chi and Ngo Ken Hui, *An Alternative Method of Image Compression and Restoration*, School of Electrical and Electronic Engineering, Nanyang Technological University, Singapore, Report, March 1999.

C.N. Chen and D.I. Hoult, *Biomedical Magnetic Resonance Technology*, Adam-Hilger, Bristol, 1989.

S. Choi, Design of an adaptive antenna array for tracking the source of maximum power and its application to CDMA mobile communications, *IEEE Trans. Antennas Propagation*, **45**, 1393–1404, 1997.

R.T. Compton, *Adaptive Antennas: Concepts and Applications*, Prentice-Hall, Englewood Cliffs, 1988.

H. Cox, R. Zeskind and M.M. Owen, Robust adaptive beamforming, *IEEE Trans. Acoustics*, **35**, 1987.

D.J. Daniels, *Surface Penetrating Radar*, Institution of Electrical Engineers, London, 1996.

S. Drabowitch, A. Papiernik, H. Griffiths, J. Encinas and B.L. Smith, *Modern Antennas*, Chapman and Hall, London, 1998.

T.C. Farrar and E.D. Becker, *Pulse and Fourier Transform NMR*, Academic Press, London, 1971.

R.P. Feynman, R.O. Leighton and M. Sands, *Lectures on Physics*, Vol. 2, Addison-Wesley, London, 1964.

P.J. Fitch, *Synthetic Aperture Radar*, Springer-Verlag, Berlin, 1988.

Goh Seow Hee and Tan Pek Hua, *A Signal Processor for Tracking and Imaging a Landing Aircraft*, School of Electrical and Electronic Engineering, Nanyang Technological University, Singapore, Report, April 1998.

G.C. Hess, *Land-Mobile Radio Engineering*, Artech House, Boston, 1993.

Y. Ilo and M.A. Ingran, Design of partially adaptive arrays using the singular value decomposition, *IEEE Trans. Antennas Propagation*, **45**, 1997.

R. Hohno and H. Imai, Combination of an adaptive array antenna and a cancellor of interference for DS-SSMA system, *IEEE Journal Comms*, **8**, 1990.

P.R.P. Hoole (Editor), *Electromagnetic Imaging in Science and Medicine: With Wavelet Applications*, WIT Computational Mechanics Publishers, Southampton, 2000.

S.R.H. Hoole, Computer Aided Analysis and Design of Electromagnetic Devices, Elsevier, New York, 1987.

S.R.H. Hoole (Editor), *Finite Elements, Electromagnetics and Design*, Elsevier, New York, 1995.

S.R.H. Hoole and P.R.P. Hoole, *A Modern Short Course in Engineering Electromagnetics*, Oxford University Press, New York, 1996.

W.C. Jakes, A comparison of specific space diversity techniques for reduction of fast fading in UHF mobile radio systems, *IEEE Trans. Vehicular Technology*, **20**, 81–92, 1971.

S.S. Jeng, Experimental evaluation of smart antenna system performance for wireless communications, *IEEE Trans. Antennas Propagation*, **46**, 749–757, 1998.

J.D. Kraus, *Antennas*, McGraw-Hill, New York, 1988.

E. Krestal (Editor), Imaging Systems for Medical Diagnostics, Siemens, 1990.

H.N. Kritikos and D.L. Jaggard, *Recent Advances in Electromagnetic Theory*, Springer, New York, 1990.

W.C.Y. Lee, Mobile Communication Engineering Theory and Applications, McGraw-Hill, New York, 1998.

J.C. Liberti and T.S. Rappaport, *Smart Antennas for Wireless Communications*, Prentice-Hall, Englewood Cliffs, 1999.

T.M. Lillesand and R.W. Kiefler, *Remote Sensing and Image Interpretation*, 3rd edn, John Wiley, New York, 1994.

J. Litva and T.K.Y. Lo, *Digital Beamforming in Wireless Communications*, Artech House, Boston, 1996.

Chun Loo and Norman Second, Computer model for fading channel with application to digital transmission, *IEEE Trans. Vehicular Technology*, **40** (4), 700–707, 1991.

M.T. Ma, Theory and Application of Antenna Arrays, John Wiley, New York, 1974.

A. Mehrotra, *GSM System Engineering*, Artech House, Boston, 1997.

R.A. Monzingo and T.W. Miller, *Introduction to Adaptive Arrays*, John Wiley, New York, 1980.

H. Mott, Antennas for Radar and Communications: A Polarimetric Approach, John Wiley, New York, 1992.

T.S. Naveendra and P.R.P. Hoole, Near field computation for medical imaging, *IEEE Trans. Magnetics*, **35**, 1999.

T.S. Naveendra and P.R.P. Hoole, A generalized finite sized dipole model for radar and medical imaging. Part I: Near field formulation for radar imaging, in J.A. Kong (Editor), *Progress in Electromagnetics Research*, PIER 24, pp. 201–225, 1999.

T.S. Naveendra and P.R.P. Hoole, A generalized finite sized dipole model for radar and medical imaging. Part II: Near field formulation for magnetic resonance imaging, in J.A. Kong (Editor), *Progress in Electromagnetics Research*, PIER 24, pp. 227–256, 1999.

T.S. Naveendra, *Near Electromagnetic Fields in Medical and Radar Imaging*, School of Electrical and Electronic Engineering, Nanyang Technological University, Singapore, Thesis, 1999.

Ng Joo Seng and Seow Chee Kiat, *Near Range Communication and Radar Network for Traffic Guidance and Control of Airport Surface*, School of Electrical and Electronic Engineering, Nanyang Technological University, Singapore, Report, 1998.

Oh Kok Leong and Ng Kim Chong, *Beamforming with Position and Velocity Estimation in Cellular Communication*, School of Electrical and Electronic Engineering, Nanyang Technological University, Singapore, Report, April 1999.

A.F. Naguib, A. Paulraj and T. Kailath, Capacity improvement with base station antenna arrays in cellular CDMA, *IEEE Trans. Vehicular Technology*, **43**, 1994.

J.D. Parsons, M. Henze, P.A. Ratliff and M.J. Withers, Diversity techniques for mobile radio reception, *IEEE Trans. Vehicular Technology*, **25**, 75–84, 1976.

A.J. Paulraj and C.B. Papadias, Space–time processing for wireless communications, *IEEE Signal Processing Magazine*, pp. 33–41, November 1997.

L.H. Randy, Phase-only adaptive nulling with a genetic algorithm, *IEEE Trans. Antennas Propagation*, **45**, 1997.

T.S. Rappaport, Smart Antennas, Adaptive Arrays, Algorithms, and Wireless Position Location: Selected Readings, IEEE Press, London, 1998.

B. Razavi, *RF Microelectronics*, Prentice-Hall, Englewood Cliffs, 1998.

E. Roubine and J.C. Bolomey, *Antennas*, Vols 1 and 2, North Oxford Academic, 1987.

A.W. Rudge, K. Milne, A.D. Olver and P. Knight, *The Handbook of Antenna Design*, Peter Peregrinus, London, 1986.

G. Sebastiani and P. Barone, Mathematical Principles of Basic Magnetic Resonance Imaging in Medicine, *J. Signal Processing*, 1991.

M. Soumekh, *Fourier Array Imaging*, Prentice-Hall, Englewood Cliffs, 1994.

G.W. Stimson, *Introduction to Airborne Radar*, Hughes Aircraft Co., El Sugundo, California, 1983.

G.L. Stuber, *Mobile Wireless Communications*, Kluwer, Dordrecht, 1996.

W.L. Stutzman and G.A. Thiele, *Antenna Theory and Design*, John Wiley, New York, 1998.

S. Taira, M. Tanka and S. Ohmori, High gain airborne antenna for satellite communications, *IEEE Trans. Aerospace and Electronic Systems*, **27** (2), 354–359, 1991.

R. Vijayan, A model for analyzing handoff algorithms, *IEEE Trans. Vehicular Technology*, **42**, 351–356, 1993.

X. Wang, P.R.P. Hoole and E. Gunawan, An electromagnetic-time delay method for determining the positions and velocities of mobile stations in GSM network, in J.A. Kong (Editor), *Progress in Electromagnetics Research*, PIER 23, pp. 165–186, 1999.

W. Wharton, S. Metcalfe and G.C. Platts, *Broadcast Transmission Engineering Practice*, Focal Press, UK, 1992.

B. Widrow and S. Stearns, *Adaptive Signal Processing*, Prentice-Hall, Englewood Cliffs, 1985.

S.J. Yu and J.H. Lee, Design of two-dimensional rectangular array beamformers with partial adaptivity, *IEEE Trans. Antennas Propagation*, **45** (1), 1997.

WITPRESS

Electromagnetic Imaging in Science and Medicine

With Wavelet Applications

Editor: **P.R.P. HOOLE**, *Nanyang Technological University, Singapore*

Four highly advanced technologies – biomedical imaging, radar, navigation and remote sensing – have brought the importance of electromagnetic imaging into prominence. Because of the relatively independent development of these technologies, however, there has so far been little attempt to put together a unified approach to the electromagnetic imaging principles that are applicable to all four disciplines. This volume fills this gap. It introduces research and development engineers, graduate students and senior undergraduate students to the basic principles and techniques involved in electromagnetic image reconstruction and image processing.
An attractive feature of the book is that in addition to covering the fundamental science behind imaging and the algorithms most commonly used, it also provides technological implementation examples. All the concepts and applications of electromagnetic imaging considered essential are discussed, while the programs listed will be of help to beginners who require a starting point for reconstructing images, as well as for manipulating and processing these. Signal processing is addressed at the levels of both images and the time domain electromagnetic signals and basic tools available for processing images are described. The application of wavelets and wavelet transforms to electromagnetic imaging systems is presented as a general approach to processing and understanding electromagnetic images.
Providing wide-ranging, yet detailed, coverage, **Electromagnetic Imaging in Science and Medicine** features the following chapters: Introduction; Fundamentals of Image Processing; Signals in Science and Medicine; Fundamentals of Image Reconstruction; Mathematics of Signals and Image Reconstruction; Wavelet Transforms for Electromagnetic Images; Magnetic Resonance Signal Analysis Using Wavelets; Vision-Based Underwater Cable Tracking; Three-Dimensional Image Reconstruction; References.

ISBN: 1-85312-770-1   2000   288pp
£95.00/US$155.00

Software for Electrical Engineering

Analysis and Design V

Edited: Wessex Institute of Technology, UK

Focusing on the design, construction, evaluation and use of software systems, this book contains papers from the Fifth International Conference on Software for Electrical Engineering Analysis and Design (ELECTROSOFT).
Although they address common goals, the software topics covered are very broad, spanning numerical algorithms, data structures, aspects of programming methodology and user interface designs. The application areas represented are equally diverse, ranging from classical electromechanics to quantum electronics and even virtual reality systems. Specific topics highlighted include: Interfaces; Package Design; Packages; Software Engineering; Numerical Methods; Symbolic Computation; Parallel Computation; and Simulation.
Series: Software Studies, Vol 3

ISBN: 1-85312-866-X   2001   apx 224pp   apx £85.00/US$132.00

WIT Press is a major publisher of engineering research. The company prides itself on producing books by leading researchers and scientists at the cutting edge of their specialities, thus enabling readers to remain at the forefront of scientific developments. Our list presently includes monographs, edited volumes, books on disk, and software in areas such as: Acoustics, Advanced Computing, Architecture and Structures, Biomedicine, Boundary Elements, Earthquake Engineering, Environmental Engineering, Fluid Mechanics, Fracture Mechanics, Heat Transfer, Marine and Offshore Engineering and

WITPress
Ashurst Lodge, Ashurst, Southampton, SO40 7AA, UK.
Tel: 44 (0) 238 029 3223
Fax: 44 (0) 238 029 2853
E-Mail: witpress@witpress

Management Information Systems

Editors: **C.A. BREBBIA**, *Wessex Institute of Technology, UK and* **P. PASCOLO**, *Università degli Studi di Udine, Italy*

Management Information Systems (MIS) are rapidly finding applications in many areas including environmental conservation, economic planning, resource integration, cartography, urban planning, risk assessment, pollution control, and transport management systems.

Representing the state-of-the-art in MIS, the contributors to this book reflect on the ways in which this technology plays an active role in linking together economic development and environmental conservation planning through its own unique characteristics. They also review its growing use in various fields of application.

The papers featured were originally presented at the major international Management Information Systems Conference and span the following broad range of topics: Environmental Management; Economic Development Management; Applications of GIS; Modelling Issues; MIS in Training and Education; The World Wide Web; Applications of MIS; Remote Sensing; Integrated Modelling and Management; and Hydro-Informatics and Geo-Informatics.

Series: Management Information Systems, Vol 1

ISBN: 1-85312-815-5   2000   516pp   £169.00/US$259.00

Applied Virtual Instrumentation

R. BAICAN, *Adam Opel AG, Germany and* **D.S. NECSULESCU**, *University of Ottawa, Canada*

This book covers the fundamental knowledge needed for interfacing sensors with a PC using the new framework of virtual instrumentation. The authors focus on the knowledge required by a non-professional to develop a modern monitoring system, i.e. to connect sensors to a PC, to drive GPIB instruments, condition their signals when required, and store and process the data using digital signal processing subroutines available in commercial virtual instrumentation packages.

Specifically written for senior undergraduates in engineering and science, as well as practising engineers and researchers with an interest in using computer-based instrumentation, but with limited knowledge of computer hardware and monitoring software, the text contains numerous numerical and programming examples. An accompanying CD-ROM features data and program files that can be used to facilitate the preparation of experimental demonstrations.

Including basic topics and background information, the text is divided into the following chapters: Introduction; LabVIEW and HP VEE Packages; Signals and Measuring Configuration; Sensors; Signal Acquisition; Sensors Characteristics Measurement; Instrument Control; Signal Generation, Analysis and Processing.

ISBN: 1-85312-800-7   2000
296pp + CD-ROM   £145.00/US$225.00

All prices correct at time of going to press but subject to change.
WIT Press books are available through your bookseller or direct from the publisher.

Save 10% when you order from our encrypted ordering service on the web using your credit card.